Limit Theorems on Large Deviations for Markov Stochastic Processes

Mathematics and Its Applications (*Soviet Series*)

Volume 38

A. D. WENTZELL
Dept. of Mathematics, Moscow State University, Moscow, U.S.S.R.

Limit Theorems on Large Deviations for Markov Stochastic Processes

Kluwer Academic Publishers
Dordrecht / Boston / London

Library of Congress Cataloging-in-Publication Data

Ventt͡sel', A. D.
[Predel'nye teoremy o bol'shikh uklonenii͡akh dli͡a markovskikh sluchaĭnykh prot͡sessov. English]
Limit theorems on large deviations for Markov stochastic processes / by A.D. Wentzell.
p. cm. -- (Mathematics and its applications (Soviet series))
Translation of: Predel'nye teoremy of bol'shikh uklonenii͡akh dli͡a markovskikh sluchaĭnykh prot͡sessov.
Bibliography: p.
Includes index.
ISBN 0-7923-0143-9
1. Markov processes. 2. Limit theorems (Probability theory) I. Title. II. Series: Mathematics and its applications (Kluwer Academic Publishers). Soviet series.
QA274.7.V4613 1989
519.2'33--dc19 89-2800

ISBN 0-7923-0143-9

Published by Kluwer Academic Publishers,
P.O. Box 17, 3300 AA Dordrecht, The Netherlands.

Kluwer Academic Publishers incorporates the publishing programmes of D. Reidel, Martinus Nijhoff, Dr W. Junk and MTP Press.

Sold and distributed in the U.S.A. and Canada
by Kluwer Academic Publishers,
101 Philip Drive, Norwell, MA 02061, U.S.A.

In all other countries, sold and distributed
by Kluwer Academic Publishers Group,
P.O. Box 322, 3300 AH Dordrecht, The Netherlands.

Original title:
Предельные теоремы о больших уклонениях для Марковских случайных процессов

Published by Nauka Publishers, Moscow, © 1986

Printed on acid-free paper

Printed in the Netherlands

SERIES EDITOR'S PREFACE

'Et moi, ..., si j'avait su comment en revenir, je n'y serais point allé.'

Jules Verne

The series is divergent; therefore we may be able to do something with it.

O. Heaviside

One service mathematics has rendered the human race. It has put common sense back where it belongs, on the topmost shelf next to the dusty canister labelled 'discarded non-sense'.

Eric T. Bell

Mathematics is a tool for thought. A highly necessary tool in a world where both feedback and non-linearities abound. Similarly, all kinds of parts of mathematics serve as tools for other parts and for other sciences.

Applying a simple rewriting rule to the quote on the right above one finds such statements as: 'One service topology has rendered mathematical physics ...'; 'One service logic has rendered computer science ...'; 'One service category theory has rendered mathematics ...'. All arguably true. And all statements obtainable this way form part of the raison d'être of this series.

This series, *Mathematics and Its Applications*, started in 1977. Now that over one hundred volumes have appeared it seems opportune to reexamine its scope. At the time I wrote

> "Growing specialization and diversification have brought a host of monographs and textbooks on increasingly specialized topics. However, the 'tree' of knowledge of mathematics and related fields does not grow only by putting forth new branches. It also happens, quite often in fact, that branches which were thought to be completely disparate are suddenly seen to be related. Further, the kind and level of sophistication of mathematics applied in various sciences has changed drastically in recent years: measure theory is used (non-trivially) in regional and theoretical economics; algebraic geometry interacts with physics; the Minkowsky lemma, coding theory and the structure of water meet one another in packing and covering theory; quantum fields, crystal defects and mathematical programming profit from homotopy theory; Lie algebras are relevant to filtering; and prediction and electrical engineering can use Stein spaces. And in addition to this there are such new emerging subdisciplines as 'experimental mathematics', 'CFD', 'completely integrable systems', 'chaos, synergetics and large-scale order', which are almost impossible to fit into the existing classification schemes. They draw upon widely different sections of mathematics."

By and large, all this still applies today. It is still true that at first sight mathematics seems rather fragmented and that to find, see, and exploit the deeper underlying interrelations more effort is needed and so are books that can help mathematicians and scientists do so. Accordingly MIA will continue to try to make such books available.

If anything, the description I gave in 1977 is now an understatement. To the examples of interaction areas one should add string theory where Riemann surfaces, algebraic geometry, modular functions, knots, quantum field theory, Kac-Moody algebras, monstrous moonshine (and more) all come together. And to the examples of things which can be usefully applied let me add the topic 'finite geometry'; a combination of words which sounds like it might not even exist, let alone be applicable. And yet it is being applied: to statistics via designs, to radar/sonar detection arrays (via finite projective planes), and to bus connections of VLSI chips (via difference sets). There seems to be no part of (so-called pure) mathematics that is not in immediate danger of being applied. And, accordingly, the applied mathematician needs to be aware of much more. Besides analysis and numerics, the traditional workhorses, he may need all kinds of combinatorics, algebra, probability, and so on.

In addition, the applied scientist needs to cope increasingly with the nonlinear world and the extra mathematical sophistication that this requires. For that is where the rewards are. Linear models are honest and a bit sad and depressing: proportional efforts and results. It is in the non-linear world that infinitesimal inputs may result in macroscopic outputs (or vice versa). To appreciate what I am hinting at: if electronics were linear we would have no fun with transistors and computers; we would have no TV; in fact you would not be reading these lines.

There is also no safety in ignoring such outlandish things as nonstandard analysis, superspace and anticommuting integration, p-adic and ultrametric space. All three have applications in both electrical engineering and physics. Once, complex numbers were equally outlandish, but they frequently proved the shortest path between 'real' results. Similarly, the first two topics named have already provided a number of 'wormhole' paths. There is no telling where all this is leading - fortunately.

Thus the original scope of the series, which for various (sound) reasons now comprises five subseries: white (Japan), yellow (China), red (USSR), blue (Eastern Europe), and green (everything else), still applies. It has been enlarged a bit to include books treating of the tools from one subdiscipline which are used in others. Thus the series still aims at books dealing with:

- a central concept which plays an important role in several different mathematical and/or scientific specialization areas;
- new applications of the results and ideas from one area of scientific endeavour into another;
- influences which the results, problems and concepts of one field of enquiry have, and have had, on the development of another.

Limit theorems for sums of independent random variables (and generalizations) form a large chapter of probability theory; they also constitute an absolutely indispensable cornerstone for statistics.

Relatively recently it has become clear, not least thanks to the work of the author of the present volume, that far more is possible in the way of limit theorems and large deviations and that a great deal can be done in the setting of general stochastic (Markov) processes. Of course, new phenomena appear compared with the classical case. Moreover, these generalized large deviation limit theorems have a multitude of important applications.

The setting is quite general; the results are striking and applicable and have not previously been treated systematically in the monographic literature; the author is a most well-known expert in the field. All in all a book that I welcome in this series with great enthusiasm.

The shortest path between two truths in the real domain passes through the complex domain.

J. Hadamard

La physique ne nous donne pas seulement l'occasion de résoudre des problèmes ... elle nous fait pressentir la solution.

H. Poincaré

Never lend books, for no one ever returns them; the only books I have in my library are books that other folk have lent me.

Anatole France

The function of an expert is not to be more right than other people, but to be wrong for more sophisticated reasons.

David Butler

Amsterdam, 25 April 1990

Michiel Hazewinkel

TABLE OF CONTENTS

NOTATIONS

χ_A	indicator of a set A;
$\overline{A}$	complement of A;
$[A]$	closure of A;
(A)	set of interior points of A;
$A_{+\delta}$	δ-neighbourhood of A;
$A_{-\delta}$	set of points of A whose distances from $\overline{A}$ are greater than δ;
$a \wedge b,\ a \vee b$	minimum, maximum of two numbers a, b;
$[\ ]$	a) integral part; b) just brackets;
$f(t-)$	left limit of a function f at a point t;
zu	scalar product of z and u;
$G(z) \leftrightarrow H(u)$	notation for the Legendre transformation;
$zA,\ Au$	notation for two mutually adjoint linear operators;
$\theta \to,\ \lim_{\theta \to}$	notations for the limit with respect to a filter;
(W, ψ)	a chart on a manifold;
$\dfrac{\partial f}{\partial x^i}, \dfrac{\partial^2 f}{\partial x^i \partial x^j}$	partial derivatives of the image of a function on a chart of a manifold
$TX_x,\ T^*X_x$	tangent and cotangent spaces at a point x of a manifold X;
A_x	the mapping of the tangent space TX_x onto R^r giving the coordinates in a given local coordinate system;
$\boldsymbol{C} = \boldsymbol{C}[0, T]$	space of continuous functions on $[0, T]$;
$\boldsymbol{D} = \boldsymbol{D}[0, T]$	space of right continuous functions on $[0, T]$ without discontinuities of the second order;
$\boldsymbol{C}^1[0, T]$	space of continuously differentiable functions on $[0, T]$;
$\boldsymbol{W}^{1,2}[0, T]$	space of absolutely continuous functions on $[0, T]$ with square integrable derivatives;
$\boldsymbol{C}_{x_0}, \boldsymbol{D}_{x_0}, \boldsymbol{C}^1_{x_0}[0, T], \boldsymbol{W}^{1,2}_{x_0}[0, T]$	subspaces of these spaces, consisting of the functions equal to x_0 at $t = 0$;
$\mathfrak{T}$	the ordered set of points $t \in [0, T]$ and $t-$, $t \in (0, T]$;
$\boldsymbol{C}(\mathfrak{T})$	the space of continuous functions on $\mathfrak{T}$;
$\|\cdot\|$	the supremum norm;
ρ	a metric in an arbitrary function space;

$\rho_{0,T}$	the metric of the least upper bound of the distance in the spaces $\boldsymbol{C}$, $\boldsymbol{D}$;
$\rho_{0,T\wedge\tau_B\wedge\tau_V}$	the semi-metric of the least upper bound of the distance over $[0,T]\cap[0,\tau_B]\cap[0,\tau_V)$ in the space $\boldsymbol{D}\,[0,T]$;
$\rho_{\{0,T\}}$	the metric in $\boldsymbol{D}\,[0,T]$ defined as $\rho_{0,T}$ for functions ϕ, ψ with $\phi(0)=\psi(0)$, $\phi(T)=\psi(T)$ and as $+\infty$ otherwise;
$\rho_{\{t_m,t_{m+1}\}}$, $\rho_{\{t_0,t_1,\ldots,t_n\}}$	same with $\phi(t_i)=\psi(t_i)$;
$\mathcal{B}_X$	σ-algebra of Borel subsets of X;
$\mathcal{B}^{[0,T]}(\boldsymbol{X})$	the σ-algebra in a function space $\boldsymbol{X}$ generated by its cylinder subsets;
$\mathcal{F}_{[t_0,t]}$	the σ-algebra in the space Ω generated by the values of a stochastic process $\xi(s)$, $s\in[t_0,t]$;
$\mathcal{F}_{t_0t}$	the same σ-algebra completed with respect to the suitable probability measure and extended by continuity on the right;
$\mathcal{P}$	σ-algebra of predictable sets;
μ_ξ	distribution of a stochastic process ξ in a function space;
P^z_{0,x_0}	probability measure obtained as the result of the generalized Cramér's transformation;
M, $\mathsf{M}^\theta_{t,x}$, M^z_{0,x_0}	expectations with respect to the probability measures P, $\mathsf{P}^\theta_{t,x}$, P^z_{0,x_0}, etc.;
$\mathsf{M}(A;\xi)$	notation for $\int_A \xi(\omega)\,\mathsf{P}(d\omega)$;
$\tilde{\eta}(t)$	compensator of a random function $\eta(t)$;
$\langle\eta,\zeta\rangle(t)$	bicompensator of random functions $\eta(t)$, $\zeta(t)$;
$\eta^c(t)$	the continuous part of a local martingale;
$\eta^d(t)$	the purely discontinuous part of a local martingale;
$\tilde{\eta}^z(t)$, $\langle\eta,\zeta\rangle^z(t)$	compensator, bicompensator with respect to the transformed probability measure P^z_{0,x_0};

$\sum_{[t_0, t)} z\Delta\eta$	discrete analogue of the stochastic integral;
$\sum_{t_0, T}^{k} (V)$	sum over the jumps of a stochastic process;
$\lambda_{t, x}(dy)$	(a) the Lévy measure corresponding to the jumps of a stochastic process; (b) the limiting measure for $g(\theta)^{-1}\lambda^{\theta}_{t, x'}(dy)$ as $\theta \to, x' \to x$ (Chapter 6);
$\mathfrak{A}$	compensating operator of a Markov process;
A_t	generator of a Markov process;
$\mathfrak{A}^z$	the compensating operator of a process after the use of the generalized Cramér's transformation;
$\mathfrak{A}_V$	the operator used in the representation of the compensator of the random function $f(t, \xi_V(t))$;
$w(t)$	Wiener process;
$\xi^{n, z}(t)$, $\xi^{n, x}(t)$	stochastic processes constructed starting from sums of independent random variables X_i: $\xi^{n, z}(t) = (X_1 + \ldots + X_{[nt]}) / z$, $\xi^{n, x}(t) = (X_1 + \ldots + X_{[nt]}) / x\sigma\sqrt{n}$;
$\xi^{\tau, h}(t)$	a process with frequent small jumps (see § 4.2);
$\eta^{h, \beta}_{x_0}(t)$, $\zeta^{h, \beta}_{x_0}(t)$	the process $\xi^{h, h}(t)$ subjected to a transformation that "straightens" the most probable paths, and expanded β times;
$\xi_V(t)$	the process $\xi(t)$ stopped at the last time before its first jump not belonging to a set V;
τ_B	first exit time from a set B;
τ_V	time of the first jump not belonging to V;
$G(t, x; z)$	cumulant of a Markov process;
$G_V(t, x; z)$	truncated cumulant;
$G_0(t, x; z)$	the function that is the limit of the cumulant (the truncated cumulant), transformed in a suitable way;
$G_*(t, x; z)$	the function used in the expression of the cumulant of the processes of a given family;

$\overline{G}(z)$, $\overline{G}_V(z)$, $\overline{G}_0(z)$ functions majorizing the corresponding functions G, G_V and G_0 of the arguments t, x, z;

$H(t, x; u)$, $H_V(t, x; u)$, $H_0(t, x; u)$, $\underline{H}(u)$, $\underline{H}_0(u)$ Legendre transforms of the corresponding functions $G(t, x; z), \ldots, \overline{G}_0(z)$;

$\overline{U}$ closure of the set of all points u where the function $\underline{H}(u)$ or $\underline{H}_0(u)$ is finite;

$k(\theta)$ normalizing coefficient;

$S(\phi)$ normalized action functional;

$I(\phi)$ action functional for an individual stochastic process;

$\Phi(s)$, $\Phi(i)$, $\tilde{\Phi}(i)$, ... the set of functions where the corresponding functional takes values not exceeding the given number;

$I_V(\phi)$ truncated action functional;

$S_{T_1, T_2}(\phi)$, $I_{T_1, T_2}(\phi)$, ... functionals associated with Markov processes on a time interval $[T_1, T_2]$;

$\Phi_{x;\,[T_1, T_2]}(s)$, $\Phi_{x;\,[T_1, T_2]}(i)$, $\tilde{\Phi}_{x;\,[T_1, T_2]}(i)$, ... sets of functions $\phi(t)$ on the interval $[T_1, T_2]$ such that $\phi(T_1) = x$ and the corresponding functional does not exceed the given number;

$\pi(0, t)$ (for $t = T$) density of the generalized Cramér's transformation;

$\pi_{B, V}(0, t)$ (for $t = T$) density of the truncated Cramér's transformation;

$g(\theta)$ the infinitely small function giving the order of $\lambda^{\theta}_{t, x}(dy)$ as $\theta \to$ (Chapter 6);

$x_0 \, {}^{t_1}x_1 \ldots {}^{t_k}x_k(t)$ the function taking values x_i for $t_i \le t < t_{i+1}$, x_k for $t_k \le t \le T$;

$B^k_{x_0}$ set of all functions $x_0 \, {}^{t_1}x_1 \ldots {}^{t_k}x_k$ with given x_0 and k;

$E^k_{x_0}$ the set $\{(t_1, x_1, \ldots, t_k, x_k)\colon 0 < t_1 < \ldots < t_k \le T,\ x_i \ne x_{i-1},\ 1 \le i \le k\}$;

$X_{x_0, k}$ the mapping taking a point $(t_1, x_1, \ldots, t_k, x_k) \in E^k_{x_0}$ to the function $x_0 \, {}^{t_1}x_1 \ldots {}^{t_k}x_k \in \boldsymbol{D}_{x_0}$;

$\mu^k_{x_0}$	the measure on $E^k_{x_0}$ defined by $\mu^k_{x_0}(dt_1\, dx_1 \ldots dt_k\, dx_k) = dt_1 \lambda_{t_1, x_0}(dx_1) \ldots dt_k \lambda_{t_k, \xi_{k-1}}(dx_k)$;
$\tau^\varepsilon(s)$	time of the first jump of size greater than ε after s;
τ^ε_i	time of the i-th jump of size greater than ε;
ν^ε	number of jumps of size greater than ε on the interval from 0 to T.

ACKNOWLEDGEMENTS

The author is grateful to the members of the Laboratory of Computational Methods of the Department of Mechanics and Mathematics of Moscow State University for their kind help in preparing the manuscript of the translation of this book.

INTRODUCTION

0.1. Problems on large deviations for stochastic processes

In recent decades a new branch of probability theory has been developing intensively, namely, limit theorems for stochastic processes. As compared to classical limit theorems for sums of independent random variables, the generalizations are going here in two directions simultaneously. First, instead of sums of independent variables one considers stochastic processes belonging to certain broad classes. Secondly, instead of the distribution of a single sum — the distribution of the value of a stochastic process at one (time) point — or the joint distribution of the values of a process at a finite number of points, one considers distributions in an infinite-dimensional function space. For stochastic processes constructed, starting from sums of independent random variables, this is the same as considering the joint distribution of an unboundedly increasing number of sums.

Let us introduce the notions concerning distributions in function spaces. Let $\xi(t)$, $t \in [0, T]$, be a stochastic process taking values in a measurable space $(X, \mathfrak{B})$. Let $\boldsymbol{X}$ be a space consisting of functions $x: [0, T] \to X$. Let $\mathfrak{B}^{[0, T]}(\boldsymbol{X})$ be the σ-algebra of subsets of $\boldsymbol{X}$, generated by all cylinder sets

$$\{x \in \boldsymbol{X}: (x(t_1), \ldots, x(t_n)) \in C\}, \quad t_i \in [0, T], \quad C \in \mathfrak{B}^n.$$

Measurability in the function space $\boldsymbol{X}$ will always be understood as measurability with respect to $\mathfrak{B}^{[0, T]}(\boldsymbol{X})$. If all sample functions ξ of the process $\xi(t)$ belong to $\boldsymbol{X}$, the distribution μ_ξ of this stochastic process in the function space $\boldsymbol{X}$ is, by definition, the measure whose value on a measurable set A is defined by $\mu_\xi(A) = \mathsf{P}\{\xi \in A\}$.

Limit theorems are usually formulated in the case when in the function space $\boldsymbol{X}$ a metric $\rho(x, y)$ is given. We do not assume that the basic σ-algebra $\mathfrak{B}^{[0, T]}(\boldsymbol{X})$ in this space coincides with the σ-algebra generated by the metric (the Borel σ-algebra); we assume only that the metric ρ is measurable and that for any measurable set A the distance $\rho(x, A)$ is measurable (from the measurability of the metric ρ we can deduce the measurability of every compact set, but not that of every closed set).

In this book $\boldsymbol{X}$ will be the space $\boldsymbol{D} = \boldsymbol{D}[0, T]$ of all right continuous functions with limits on the left in a metric space X (with its σ-algebra $\mathfrak{B} = \mathfrak{B}_X$ of Borel sets), the space $\boldsymbol{C} = \boldsymbol{C}[0, T]$ of continuous functions, or their subspaces $\boldsymbol{D}_{x_0}$, $\boldsymbol{C}_{x_0}$ consisting of functions taking the value x_0 at the point $t = 0$. On these spaces we will usually consider the metric

$$\rho_{0,T}(x, y) = \sup_{0 \le t \le T} \rho(x(t), y(t)),$$

and in § 5.2 we will consider the Skorohod metric. The metric $\rho_{0,T}$ and the Skorohod metric satisfy the above-introduced conditions (we must keep in mind that the σ-algebra $\mathfrak{B}^{[0,T]}(D)$ does not coincide with the Borel σ-algebra in D corresponding to the metric $\rho_{0,T}$).

Limit theorems for stochastic processes can be formulated as follows. Let there be given a *filter* on a set Θ of elements θ (i.e., in essence, a system of neighbourhoods of an ideal point; for the definition of filter and of limit with respect to a filter see Bourbaki [1], Chapter 1). Limits with respect to this filter will be denoted by $\lim_{\theta \to}$, or: $\to$ as $\theta \to$; the expression "for sufficiently far θ ..." will mean: "there exists a set B of the filter such that for all $\theta \in B$ ...". Let to every value of the parameter θ correspond a stochastic process $\xi^\theta(t)$, $0 \le t \le T$, on a probability space $(\Omega^\theta, \mathfrak{F}^\theta, \mathsf{P}^\theta)$ with sample functions belonging to the function space X. Limit theorems for the family of stochastic processes $\xi^\theta(t)$ deal with the limiting behaviour of the distribution μ_{ξ^θ} on $(X, \mathfrak{B}^{[0,T]}(X))$ as $\theta \to$.

For example, theorems on weak convergence state that

$$\int_X f(x)\, \mu_{\xi^\theta}(dx) \to \int_X f(x)\, \mu(dx)$$

as $\theta \to$ for every bounded measurable continuous functional f. Theorems of the type of the law of large numbers state that $\xi^\theta \to x$ in probability as $\theta \to$; i.e. that

$$\mathsf{P}^\theta\{\rho(\xi^\theta, x) \ge \varepsilon\} \to 0$$

as $\theta \to$ for all $\varepsilon > 0$. Here $x(t)$, $0 \le t \le T$, is a non-random function belonging to X.

Problems on large deviations for stochastic processes can be formulated in the following way. For a family of stochastic processes $\xi^\theta(t)$ let a result of the type of the law of large numbers take place: $\xi^\theta \xrightarrow{\mathsf{P}} x$ $(\theta \to)$. Problems on large deviations (of the process $\xi^\theta(t)$ from its "most probable" path $x(t)$ for far values of θ) are concerned with the limiting behaviour as $\theta \to$ of the infinitesimal probabilities $\mathsf{P}^\theta\{\xi^\theta \in A\}$ for measurable sets $A \subset X$ that are situated at a positive distance from the non-random limiting function x. Problems concerning asymptotics as $\theta \to$ of expectations of the form $\mathsf{M}^\theta f^\theta(\xi^\theta)$ also belong to large deviation problems if the main part of such expectations for far values of θ is due to the low-probability values of ξ^θ.

Problems on large deviations for sums of independent random variables are included in this pattern in the following way. Let $X_1, X_2, \ldots, X_n, \ldots$ be independent, identically distributed random variables; let the random variable $\zeta_n = (X_1 + \ldots + X_n - A_n) / B_n$ have a limiting distribution as $n \to \infty$. Integral limit theorems on large deviations for sums of the X_i deal with the limiting behaviour of the probability $\mathsf{P}\{\zeta_n > x\}$ as $n \to \infty$, $x \to \infty$, with certain relations between their rates of tending to infinity. So, Theorem 1 of Cramér [1] deals with the case when x tends to infinity slower than $\sqrt{n}$ (we assume that $\mathsf{M}X_i = 0$, $\mathsf{D}X_i = \sigma^2 < \infty$, so that $A_n = 0$, $B_n = \sigma\sqrt{n}$, and that the exponential moments $\mathsf{M}e^{zX_i}$ are finite for sufficiently small z). Here the parameter θ is the pair $(n, x) \in \{1, 2, \ldots\} \times (0, \infty)$. The filter in question is that corresponding to the convergence $x \to \infty$, $x = o(\sqrt{n})$ ($n \to \infty$ follows from it); that is, the filter having at its base all sets of the form $\{(n, x): x \geq x_0, x \leq \varepsilon\sqrt{n}\}$, where x_0 and ε are arbitrary positive numbers. Define the stochastic process $\xi^{n,x}(t)$ by the equation

$$\xi^{n,x}(t) = (X_1 + \ldots + X_{[nt]}) / (x\sigma\sqrt{n}), \quad t \in [0, 1].$$

As $x \to \infty$, $x = o(\sqrt{n})$ this process converges in probability to the function identically zero. For A we take the set of all functions $\boldsymbol{D}[0, 1]$, taking a value exceeding 1 at the right end of the interval $[0, 1]$. It is clear that $\mathsf{P}\{\xi^{n,x} \in A\}$ coincides with $\mathsf{P}\{\zeta_n > x\}$.

0.2. Two opposite types of behaviour of probabilities of large deviations

There are two extreme opposite types of limiting behaviour of probabilities of large deviations for sums of independent random variables. To the first type belongs the case when Cramér's condition $\mathsf{M}e^{zX_i} < \infty$ is satisfied. Here the principal part of the probability $\mathsf{P}\{\zeta_n > x\}$ is equal to the probability that ζ_n is in a small neighbourhood to the right of x, and this probability is due mainly to summands X_i approximately of the same size; the probability $\mathsf{P}\{\zeta_n > x\}$ has an exponential asymptotics. Theorems on large deviations of this type can be obtained using a transformation of distributions proposed by Cramér [1].

To the second type belongs the case when the X_i belong to the domain of attraction of a stable law with exponent $\alpha < 2$. Here the probability $\mathsf{P}\{\zeta_n > x\}$ is not equivalent to $\mathsf{P}\{\zeta_n \in (x, x(1 + \varepsilon))\}$; its main part is due to one summand which is approximately as large as the whole sum $X_1 + \dots + X_n$; here we have a power asymptotics. Results of this type for random variables attracted to a stable law were obtained by Fortus [1], Heyde [1] and Tkachuk [1].

There are some more subtle results intermediate between the two types, where the part of the probability of large deviations due to small individual summands and that due to one large summand are of the same order (see, for example, Nagaev [1]).

The two extreme opposite types of limiting behaviour of probabilities of large deviations occur for stochastic processes as well. The first type is characterized by the fact that the infinitesimal probabilities $\mathsf{P}^\theta\{\xi^\theta \in A\}$ arise mainly because of sample paths that are close to continuous or even smooth, whereas in the second type they stem from sample paths ξ^θ having one or several large jumps.

The major part of results on large deviations for stochastic processes obtained up till now belongs to the first type, and their majority does not deal with asymptotics of large deviations up to equivalence, but rather up to logarithmic equivalence:

$$\ln \mathsf{P}^\theta\{\xi^\theta \in A\} \sim \dots .$$

We will call such results *rough.*

0.3. Rough theorems on large deviations; the action functional

Such theorems for stochastic processes associated with sums of independent random variables were obtained in Borovkov [1], Varadhan [1], and, for the sample distribution function, in Sanov [1]. Logarithmic asymptotics of large deviations in these papers as well as in this book are characterized by means of a certain functional $k(\theta)$ $S(\phi)$ describing the "difficulty" for the stochastic process ξ^θ to fall into a neighbourhood of a function ϕ (a little more accurately: $\mathsf{P}^\theta\{\rho(\xi^\theta, \phi) < \delta\} \approx \exp\{-k(\theta) S(\phi)\}$ for small $\delta > 0$ and far values of θ; for the precise meaning see below). For the functional $k(\theta) S(\phi)$ we use, after Freidlin [1], Freidlin, Wentzell, [6], the term *action functional*; its individual factors depending on the parameter θ and the function ϕ are termed: $k(\theta)$, the *normalizing coefficient* (certainly, $k(\theta)$ is bound to be infinitely large as $\theta \to$); $S(\phi)$, the *normalized action functional*. Let us give the precise formulation.

Let ξ^θ be a family of stochastic processes with sample functions belonging to a function space X with metric ρ. Let $k(\theta)$ be a function taking positive real values, $k(\theta) \to +\infty$ as $\theta \to$, and let $S(\phi)$ be a functional on X with values in $[0, +\infty]$. We say that $k(\theta) S(\phi)$ is the action functional for ξ^θ as $\theta \to$ if the following conditions are satisfied:

(0) for any $s \geq 0$ the set $\Phi(s) = \{\phi : S(\phi) \leq s\}$ is compact;

(I) for any $\delta > 0$, $\gamma > 0$, $s_0 > 0$, for sufficiently far θ and for any $\phi \in \Phi(s_0)$,

$$\mathsf{P}^\theta \{\rho(\xi^\theta, \phi) < \delta\} \geq \exp\{-k(\theta)[S(\phi) + \gamma]\};$$

(II) for any $\delta > 0$, $\gamma > 0$, $s_0 > 0$, for sufficiently far θ and for any $s \leq s_0$,

$$\mathsf{P}^\theta \{\rho(\xi^\theta, \Phi(s)) \geq \delta\} \leq \exp\{-k(\theta)(s - \gamma)\}.$$

Condition (0) is not concerned with stochastic processes of our family, but with the normalized action functional $S(\phi)$ only. In the case of a complete space X this condition can be split in two: lower semicontinuity of the functional $S(\phi)$ on X (which is equivalent to closedness of $\Phi(s)$ for arbitrary s), and precompactness of $\Phi(s)$; this splitting is convenient for the verification of Condition (0).

Let us give some other definitions of the action functional, equivalent to Conditions (I) and (II) when Condition (0) is satisfied:

(I') for all open measurable $A \subseteq X$,

$$\varliminf_{\theta \to} k(\theta)^{-1} \ln \mathsf{P}^\theta \{\xi^\theta \in A\} \geq -\inf\{S(\phi): \phi \in A\};$$

(II') for all closed measurable $A \subseteq X$,

$$\varlimsup_{\theta \to} k(\theta)^{-1} \ln \mathsf{P}^\theta \{\xi^\theta \in A\} \leq -\inf\{S(\phi): \phi \in A\}.$$

We say that a set $A \subseteq X$ is *regular* (with respect to the functional S) if the greatest lower bound of S on the closure of A coincides with its greatest lower bound on the set of interior points of A. The two conditions (I), (II) or (I'), (II') can be replaced by one:

(I + II) for all regular measurable sets $A \subseteq X$,

$$\lim_{\theta \to} k(\theta)^{-1} \ln \mathsf{P}^\theta \{\xi^\theta \in A\} = -\inf\{S(\phi): \phi \in A\}.$$

Another way of describing rough asymptotics of large deviations in integral terms is:

(III) if $F(x)$ is a continuous bounded measurable functional on X, then

$$\lim_{\theta \to} k(\theta)^{-1} \ln \mathsf{M}^\theta \exp\{k(\theta) F(\xi^\theta)\} = \max\{F(\phi) - S(\phi): \phi \in X\}.$$

Conditions (I'), (II') are used in Varadhan [1] and later works by the same author; (I + II) in Borovkov [1], Mogul'skii [1], and some other papers. The equivalence of (I), (II) to the pair (I'), (II') or to Condition (I + II) given (0), is proved in Wentzell, Freidlin [6] (Theorems 3.3 and 3.4 of Chapter 3; the proof is given in case the basic σ-algebra in X is generated by the metric ρ, but it remains valid under our conditions imposed upon the metric and the basic σ-algebra). Condition (III) (and even more intricate conditions of integral type) is derived from (I') and (II') in Varadhan [1].

One easily proves the following lemma.

Lemma 0.1. *If F is a continuous bounded measurable functional on X, and if* *the difference $F - S$ attains its maximum on a unique function* $\phi_0 \in X$, *then for any* $\delta > 0$ *there exists a* $\gamma > 0$ *such that*

$$\int_{\{\rho(\xi^\theta, \phi_0) \geq \delta\}} \exp\{k(\theta) F(\xi^\theta)\} \, dP^\theta = o(\exp\{k(\theta)[F(\phi_0) - S(\phi_0) - \gamma]\})$$

as $\theta \to$.

We can take $\gamma = (F(\phi_0) - S(\phi_0) - \max\{F(\phi) - S(\phi) : \rho(\phi_0, \phi) \geq \delta\})/2$.

0.4. Survey of work on large deviations for stochastic processes

A considerable part of such results deals with families of Markov processes. We may mention papers concerned with families of processes with independent increments (including those constructed starting from an increasing number of independent variables): Borovkov [1], Varadhan [1], Mogul'skii [1], Borovkov, Mogul'skii [1]; those concerned with diffusion processes with small diffusion, with applications to different asymptotic problems (in particular, asymptotic stability problems): Freidlin, Wentzell [1] - [6]; Wentzell [1], [2], [6]; Freidlin [2], [4]; concerning diffusion processes with small diffusion reflected at the boundary: Anderson, Orey [1]; Zhivoglyadova, Freidlin [1]; with some wider classes of families of Markov processes containing both processes with independent increments and diffusion processes (but not diffusion processes with reflection): Wentzell [4], [5], [7]; Azencott, Ruget [1].

In certain related papers the asymptotics of the transition probabilities of a Markov process depending on a parameter is studied: for diffusion processes with small diffusion (or on a small time interval): Varadhan [2], [3]; Kifer [1]; Molchanov [1]; Friedman [1]; for processes with frequent small jumps (the pattern considered in Chapter 4 of this book): Maslov [1], [2] (the transition density $\rho^\theta(t, x, y)$ is logarithmically equivalent to $\exp\{-k(\theta) \min\{S_{0,t}(\phi) : \phi(0) = x, \phi(t) = y\}\}$; the precise asymptotics is related with the concrete form of dependency of the family of processes upon the parameter and is, in certain cases, expressed in a complicated way). In Mogul'skii [2] a precise asymptotics is found for the probability that a process $\xi^\theta(t)$ with independent increments passes through a strip between two curves.

In Schilder [1], Dubrovskii [1], [2] precise asymptotics of large deviations in a function space is studied, namely asymptotics of expectations of the form $M^\theta \exp\{k(\theta) F(\xi^\theta)\}$ where F is a smooth functional, or of the form $M^\theta G(\xi^\theta) \exp\{k(\theta) F(\xi^\theta)\}$ (the rough asymptotics of such expectations is given by $\exp\{k(\theta) \max_\phi [F(\phi) - S(\phi)]\}$).

In the case of non-Markov processes ξ^θ theorems on the action functional and their applications can be found in: Freidlin [1]; Wentzell [3] (stochastic processes arising from a Gaussian one multiplied by a small parameter); Grin' [1]; Nguyen Viet Phu [1], [2] (the same, with applications to various asymptotic problems). In Freidlin [3], [5] (see also Freidlin, Wentzell [6]) large deviations are studied in the case when the averaging principle is valid; in particular for stochastic processes η^θ defined by differential equations $\dot{\eta}^\theta(t) = b(\eta^\theta(t), \xi(\theta t))$, as $\theta \to \infty$, where $\xi(s)$ is a stationary stochastic process.

In Donsker, Varadhan [1] and Gärtner [1] large deviations of the random measure

$$\pi^T(\Gamma) = T^{-1}\int_0^T \chi_\Gamma(\xi(t))\,dt$$

from the limiting measure of π^T as $T \to \infty$ are studied, where $\xi(t)$ is a stochastic process. The random function π^T is not defined on an interval of the real line, but rather on a σ-algebra of sets. Important applications to theoretical physics are considered in the above-mentioned work of Donsker and Varadhan, as well as in their paper [2].

Problems on large deviations of the second type were studied only in few papers: Godovan'chuk [1], [2] (for Markov processes with a number of large jumps decreasing in a certain way) and Pinelis [1] (for families of processes with independent increments). The latter paper contains some subtle results intermediate between the first and the second type, analogous to Nagaev's [1] results for sums of independent random variables.

0.5. The scheme for obtaining rough theorems on large deviations

The first chapter contains the auxiliary material and devices allowing one to work with Markov processes. These devices are based on the important concept of the compensator of a random function and ultimately on martingales.

In Chapters 2 - 4 we obtain rough limit theorems on large deviations for families of Markov processes. The presentation is based on Wentzell [7]. In the present book we follow this scheme for obtaining rough theorems on large deviations (see Wentzell [4], [5], the first paper of [7] (Introduction); the same scheme is followed in Wentzell [3], not for Markov processes but for Gaussian processes). To a stochastic process $\xi(t)$ belonging to a certain class, irrespective of whether it belongs to some family or not, we associate an *action functional* $I(\phi)$ (§ 2.1 of the present work). The probability for the sample path ξ to be at a distance at most δ from a function ϕ is estimated from below by the expression $\exp\{-I(\phi) - R_1(\phi, \delta)\}$ (§ 2.2), and the probability that the sample path turns out to be farther than δ from the set $\Phi(i)$ of all functions ϕ with $I(\phi) \le i$ is estimated from above by an expression of the type $\exp\{-i + R_2(i, \delta)\}$ (§

2.3). For a family of processes ξ^{θ} depending in a certain way on a parameter, the corresponding functional $I^{\theta}(\phi)$ may turn out to be equivalent to $k(\theta) S(\phi)$ as $\theta \to$ (where $k(\theta) \to \infty$). If the remainder terms $R_1^{\theta}(\phi, \delta)$, $R_2^{\theta}(k(\theta) s, \delta)$ are $o(k(\theta))$ as $\theta \to$, then from this pair of estimates we obtain the pair of Conditions (I), (II) (see § 3.2 and § 3.3). Note that the action functional $k(\theta) S(\phi)$ for the family of stochastic processes and $I^{\theta}(\phi)$ for the individual process need not coincide.

Unfortunately, this simple scheme has to be made more complicate and obscure in order to extend its range of applicability. So, it proves possible and sometimes necessary to consider, instead of the action functional $I(\phi)$, another, "alien" functional $\tilde{I}(\phi)$ and to derive estimates for $\mathsf{P}\{\rho(\xi, \tilde{\Phi}(i)) \geq \delta\}$ where $\tilde{\Phi}(i) = \{\phi: \tilde{I}(\phi) \leq i\}$ (see Theorem 2.3.2); another complication is that a truncated action functional $I^V(\phi)$ is introduced, and we obtain the more complicated estimates associated with it (§ 2.4).

Some words on how we obtain the estimates associated with the action functional (for Markov processes, in this book, and for Gaussian processes in Wentzell [3]). To obtain the estimate from below for $\mathsf{P}\{\rho(\xi, \phi) < \delta\}$ we construct a functional $\pi_{\phi}(\xi)$ of sample functions of the stochastic process and consider a new probability:

$$\tilde{\mathsf{P}}(A) = \int_A \pi_{\phi}(\xi)\, d\mathsf{P}.$$

The functional π_{ϕ} is chosen such that $\tilde{\mathsf{P}}(\Omega) = 1$ and the sample paths of the process fall, with large $\tilde{\mathsf{P}}$-probability, in a neighbourhood of the function ϕ. Then we use the equality

$$\mathsf{P}\{\rho(\xi, \phi) < \delta\} = \int_{\{\rho(\xi, \phi) < \delta\}} \pi_{\phi}(\xi)^{-1}\, d\tilde{\mathsf{P}}.$$

On a part of the range of integration having a sufficiently large $\tilde{\mathsf{P}}$-probability we estimate $\pi_{\phi}(\xi)$ from above: $\pi_{\phi}(\xi) \leq \pi_{\phi}(\phi)\, e^{R_1}$, and we obtain an estimate of the type required with an action functional of the form $I(\phi) = \ln \pi_{\phi}(\phi)$.

This device has been used by Cramér [1] in connection with distributions of sums of independent random variables, not in order to obtain rough results, but precise ones. Therefore we call the above-described transformation the *generalized Cramér's transformation*. In the Markov case the functional $\pi_{\phi}(\xi)$ can be expressed by means of stochastic integrals:

$$\pi_{\phi}(\xi) = \exp\left\{\int z(t, \xi(t-))\, d\xi(t) - \int G(t, \xi(t); z(t, \xi(t)))\, dt\right\}$$

(or, in the discrete case, by means of the corresponding sums), where the function $z(t, x)$ is defined starting from a function $G(t, x; z)$ associated with the process $\xi(t)$, and from the derivative $\dot{\phi}(t)$. The (unused) arbitrariness in the choice of the function $z(t, x)$ and the functional $\pi_{\phi}(\xi)$ proves to be unnecessary in obtaining rough theorems on large deviations; but in Dubrovskii [1], [2] Cramér's transformation is applied to the study of precise results on the asymptotics of large deviations. These results will be set forth in Chapter 5. In these papers of Dubrovskii Cramér's transformation is generalized still further and this leads to some still more precise results.

The estimate from above is obtained in the following way. Starting from the random function ξ with non-smooth sample functions for which the functional I is generally infinite, we construct an auxiliary random function $l(\xi)$ (a "polygon"), and we introduce an event A such that $A \subseteq \{\rho(\xi, l(\xi)) < \delta\}$. Then we use the inequality

$$\mathsf{P}\{\rho(\xi, \Phi(i)) \geq \delta\} \leq \mathsf{P}(\overline{A}) + \mathsf{P}(A \cap \{l(\xi) \notin \Phi(i)\}). \tag{0.1}$$

The second summand, the principal one, is estimated by means of an exponential Chebyshev inequality:

$$\mathsf{P}(A \cap \{l(\xi) \notin \Phi(i)\}) = \mathsf{P}(A \cap \{I(l(\xi)) > i\}) \leq$$

$$\leq \int_A \exp\{(1-\kappa) I(l(\xi))\} \, d\mathsf{P} / \exp\{(1-\kappa) i\}.$$

The integral proves to be not exceedingly large, and its logarithm and $\kappa \cdot i$ are included into the remainder term R_2. The first term at the right side of (0.1) is estimated by means of an exponential Chebyshev inequality too.

0.6. The expression: $k(\theta) S(\phi)$ is the action functional uniformly over a specified class of initial points

Let us say something on peculiarities in connection with the fact that in the Markov case it is natural to consider processes starting at an arbitrary point of the state space (see Dynkin [1]).

First of all, the metric we will consider is that of the least upper bound of the distance; namely, for processes on the time interval $(T_1, T_2]$ we consider

$$\rho_{T_1, T_2}(\phi, \psi) = \sup_{T_1 \leq t \leq T_2} \rho(\phi(t), \psi(t)),$$

where ρ is the metric in the space X. We will also consider a normalized action functional $S(\phi)$ dependent on the time interval $[T_1, T_2]$ on which the process is considered: $S(\phi) = S_{T_1, T_2}(\phi)$. The notation $\Phi_{x; [T_1, T_2]}(s)$ is used for the set of all functions $\phi(t)$, $T_1 \le t \le T_2$, such that $\phi(T_1) = x$ and $S_{T_1, T_2}(\phi) \le s$.

Most frequently we will consider the time interval $[0, T]$.

We say that $k(\theta) S_{0,T}(\phi)$ is the *action functional* for the family of Markov processes $(\xi^\theta(t), \mathsf{P}^\theta_{t,x})$, $t \in [0, T]$, as $\theta \to$, *uniformly* as the initial point changes over the sets of a class $\mathfrak{M}$ of subsets of the state space X if

(0_K) the functional $S_{0,T}$ is lower semicontinuous, and the union of the $\Phi_{x;[0,T]}(s)$ over all $x \in K$ is compact for every compact $K \subseteq X$;

$(\mathrm{I}_{\mathfrak{M}})$ for any $\delta > 0$, $\gamma > 0$, $s_0 > 0$, and $M \in \mathfrak{M}$, for sufficiently far θ, for all $x \in M$, and all $\phi \in \Phi_{x;[0,T]}(s_0)$,

$$\mathsf{P}^\theta_{0,x}\{\rho_{0,T}(\xi^\theta, \phi) < \delta\} \ge \exp\{-k(\theta)[S_{0,T}(\phi) + \gamma]\};$$

$(\mathrm{II}_{\mathfrak{M}})$ for any $\delta > 0$, $\gamma > 0$, $s_0 > 0$, and $M \in \mathfrak{M}$, for sufficiently far θ, for all $s \le s_0$, and $x \in M$,

$$\mathsf{P}^\theta_{0,x}\{\rho_{0,T}(\xi^\theta, \Phi_{x;[0,T]}(s)) \ge \delta\} \le \exp\{-k(\theta)(s - \gamma)\}.$$

The uniformity of the estimates associated with the action functional over a sufficiently wide class of sets of initial points is essential for applications of large deviation theorems involving the Markov property. In the majority of cases we obtain uniformity as the initial point changes over the whole state space X; sometimes only within each compact subset of X.

0.7. Chapters 3 - 6: a survey

The author preferred not to prove a great number of different limit theorems on large deviations for Markov processes independently from one another, but rather to deduce them from certain few standard results of general nature. This rôle in the present work is played by Theorems 3.2.3, 3.2.3' and 3.3.2, 3.3.2' (Theorems 3.2.3, 3.3.2 pertain to the continuous case; 3.2.3', 3.3.2' to the discrete case).

Chapter 4 contains special results derived from the general ones of Chapter 3 (of course the results given in this chapter do not exhaust all one can deduce from the general theorems). Among these results one can single out results of the type of "not very large deviations" (§ 4.4 and § 4.5), "very large" (§ 4.3), and "super-large deviations" (§ 4.6). (For the discrimination between these cases see § 4.2; for the problem of asymptotics of the probability $\mathsf{P}\{\zeta_n > x\}$ for the normalized sum of n independent variables the three cases correspond to those of $x = o(\sqrt{n})$, of x of the same order as $\sqrt{n}$, and of x tending to infinity faster than $\sqrt{n}$.)

All these results require that an analogue of Cramér's condition of finiteness of exponential moments should be satisfied, with the exception of Theorems 4.4.2, 4.4.2', 4.4.3 concerning the case of not very large deviations. These theorems are analogous to the results on large deviations for sums of independent random variables without exponential moments, but with a sharper restriction on the rate of growth of x, than $x = o(\sqrt{n})$, depending on the rate of decrease of the "tails" of the distribution (see Ibragimov, Linnik [1], Theorem 13.1.1 which deals, of course, with precise results and not with rough ones, as in our case).

The results of Chapter 4 (from Wentzell [7]), even though they are special as compared to those of the third chapter, still are formulated in the most general possible form. This made it possible, in particular, to apply them to problems concerning convergence with probability one of stochastic approximations (see Korostelev [1], [2]).

In the fifth chapter the results of Schilder [1], Dubrovskii [1], [2] are presented; these results pertain to precise limit theorems on large deviations for Markov stochastic processes. They concern the asymptotics not of probabilities $\mathsf{P}^{\theta}_{0,x}\{\xi^{\theta} \in A\}$, but of expectations of the form

$$\mathsf{M}^{\theta}_{0,x} \exp\{k(\theta) F(\xi^{\theta})\}, \quad \mathsf{M}^{\theta}_{0,x} G(\xi^{\theta}) \exp\{k(\theta) F(\xi^{\theta})\}$$

for smooth functionals F (and less smooth G).

The sixth chapter deals with theorems concerning the large deviations occurring because the Markov process performs one or several large jumps (in terms of this introduction, results of the second type). The results presented were obtained in Godovan'chuk [1], [2]. Results of this type and their possible applications have not been given sufficient attention up till now. As an example, in this chapter we consider their application to obtaining certain new results in the classical situation of sums of independent random variables (the results were obtained in Vinogradov [1]).

0.8. Numbering

Formulas, theorems and lemmas in this book are numbered with section number (two figures) and the number within the section. Sections are subdivided into subsections of varying size. The end of a proof is marked with the sign $\Diamond$.

CHAPTER 1

GENERAL NOTIONS, NOTATION, AUXILIARY RESULTS

1.1 General notation. Legendre transformation

1.1.1. General notation

The minimum and the maximum of two numbers will be denoted by $a \wedge b$, $a \vee b$, respectively; χ_A denotes the indicator function of a set A.

The complement of a set A (in a certain space) is denoted by $\overline{A}$. In a metric space, $[A]$ denotes the closure of A; (A) denotes the set of interior points of A; $A_{+\delta}$ denotes the δ-neighbourhood of A and $A_{-\delta}$ is the set of all points of A lying at a distance greater than δ from $\overline{A}$.

The left-hand limit of a function at a point t is denoted by the value of the function at the point $t-$; the derivative with respect to t will be denoted by a dot above the letter.

We shall, as long as it does not lead to confusion in understanding, use the same notations both in the case of continuous-time stochastic processes and of processes with a discrete-time parameter taking as its values multiples of some positive τ. So, a time interval $[t_0, T]$ in the discrete case is understood as consisting of multiples of τ, from t_0 up to T.

We shall use script letters to denote systems of sets, usually σ-algebras; $\mathfrak{B}_X$ will be the σ-algebra of Borel subsets of a set X.

Elements of R^r are denoted either with the letter z, and then we write the coordinate numbers as subscripts (for example: $z(t, x) = (z_1(t, x), ..., z_r(t, x)))$, and put the notation of a linear operator in R^r to the right of the notation of the vector; or using other letters, in which case the coordinate numbers are written as superscripts ($u = (u^1, ..., u^r)$) and the notation of a linear operator stands to the left of the notation of the vector. We use the notation without brackets:

$$zu = \sum_{i=1}^{r} z_i u^i;$$

we denote an operator and its adjoint by the same letter: $(zA)\, u = z\,(Au) = zAu$.

1.1.2. Legendre transformation

If $G(z)$ is a downward convex, lower semicontinuous function of $z \in R^r$, taking values in $(-\infty, \infty]$, its *Legendre transform* is the function $H(u)$, $u \in R^r$, defined by the formula

$$H(u) = \sup_z [zu - G(z)]. \tag{1.1.1}$$

This transform is also downward convex and lower semicontinuous. (Note that a convex function can have discontinuities only at the boundary of the set on which it is finite.) In the class of downward convex and lower semicontinuous functions the Legendre transformation is, as is well known (see Rockafeller [1]), its own inverse:

$$G(z) = \sup_u [zu - H(u)]. \tag{1.1.2}$$

For this transformation we will use the notation $G(z) \leftrightarrow H(u)$.

We mention without proof some properties of convex functions and of the Legendre transformation. (One can find the properties of this transformation in Rockafellar [1].)

Lemma 1.1.1. *Let $G(z) \leftrightarrow H(u)$; let z be a fixed non-zero vector, let h be the greatest lower bound of $H(u)$ on the hyperplane $\{u: zu = d\}$, and let there be points on both sides of this hyperplane where $H(u)$ is finite. Then there exists a constant c such that $H(u) \geq czu - cd + h = czu - G(cz)$ for all u.*

Lemma 1.1.2. *Let $G_1(z) \leftrightarrow H_1(u)$, $G_2(z) \leftrightarrow H_2(u)$, $\kappa < 1$, and let γ be a real number. The inequality*

$$G_1((1-\kappa)z) \leq (1-\kappa) G_2(z) + \gamma$$

holds for all z if and only if the inequality

$$H_1(u) \geq (1-\kappa) H_2(u) - \gamma$$

holds for all u.

Further, if $G(z) < \infty$ for all z, then

$$\lim_{|u| \to \infty} H(u) / |u| = \infty.$$

If the function H is differentiable at a point u, then

$$H(u) = \nabla H(u) \cdot u - G(\nabla H(u)).$$

If, in addition, the function G is differentiable at the point $z = \nabla H(u)$, then

$$\nabla G(\nabla H(u)) = u.$$

If the function G is differentiable $k \geq 2$ times at a point z and the matrix of second order derivates, $(\partial^2 G / \partial z_i \partial z_j)$, is non-degenerate at this point, then the function H is also k times differentiable at the point $\nabla G(z)$ and

$$(\partial^2 H / \partial u^i \partial u^j)\Big|_{u = \nabla G(z)} = (\partial^2 G / \partial z_i \partial z_j)^{-1}.$$

The notation $G(t, x; z) \leftrightarrow H(t, x; u)$ will mean that the Legendre transform is applied individually for each (t, x).

1.1.3 Notations related to manifolds

We denote charts of a manifold X by (W, ψ), ψ being a one-to-one map of an open set $W \subseteq X$ onto an open subset of R^r; according to 1.1.1, the coordinates of $\psi(x)$, $x \in W$, are denoted by $\psi^i(x)$. For a smooth function f on X and for a certain choice of chart $\partial f/\partial x^i$, $\partial^2 f/\partial x^i \partial x^j$ will denote the partial derivatives of the image of the function on the chart.

By TX_x, T^*X_x we denote the tangent and cotangent spaces at a point x; elements of the cotangent space will be denoted by z, and we will use for $z \in T^*X_x$, $u \in TX_x$ the notation zu.

The Legendre transformation defined by formula (1.1.1) maps any downward convex, lower semicontinuous function on T^*X_x to a function with the same properties on TX_x.

We will use the notations zA, Au for linear operators in the case of $z \in T^*X_x$, $u \in TX_x$ as well. To every chart (W, ψ), $W \ni x$, corresponds a linear map A_x from the tangent space TX_x onto R^r (the components of the vector $A_x u$, $u \in TX_x$, are the coordinates of u in the given local coordinate system); the corresponding map from T^*X_x onto R^r is then (the adjoint of) A_x^{-1}.

We say that an atlas on the manifold satisfies Condition λ if there exists a $\lambda > 0$ such that for each point there is a chart (W, ψ) in the atlas such that W contains the λ-neighbourhood of this point and the ratio of the Euclidean distance $|\psi(y) - \psi(x)|$ to the distance $\rho(x, y)$ on the manifold is bounded by some positive constants from above and from below, uniformly with respect to all x, $u \in W$ and all charts of the atlas.

1.1.4. Notations pertaining to probability spaces

If $(\Omega, \mathfrak{F}, \mathsf{P})$ is a probability space, then for an event A and a random variable ξ we denote by $\mathsf{M}(A; \xi)$ the integral $\int_A \xi(\omega)\,\mathsf{P}(d\omega)$. If the probability P has some subscripts, we provide the expectation symbol M with the same sub- or superscripts.

We consider σ-algebras $\mathfrak{F}_{[t_0, t]}$ generated by the values of the stochastic process $\xi(s)$, $s \in [t_0, t]$; $\mathfrak{F}_{t_0 t}$ denotes the same σ-algebras completed with respect to the appropriate probability measure and extended by continuity on the right in the continuous case.

1.1.5. The metric

We consider the metric

$$\rho_{T_1,T_2}(\phi,\psi) = \sup_{T_1 \le t \le T_2} \rho(\phi(t),\psi(t))$$

in spaces of functions $\phi(t)$, $T_1 \le t \le T_2$; in the case of a Euclidean space

$$\rho_{T_1,T_2}(\phi,\psi) = \sup_{T_1 \le t \le T_2} |\phi(t)-\psi(t)|.$$

This notation will be used both in the continuous and in the discrete case.

1.2. Compensators. Lévy measures

The concept of compensator (although the terminology is not fully established yet) and its related concepts are used widely in research on stochastic processes (see, for example, Meyer [1], Liptser, Shiryayev [1]). Let us set forth everything we will be needing.

1.2.1. Compensators

Let $\mathfrak{F}_t$, $t \in [t_0, T]$, be a non-decreasing family of σ-algebras in a probability space $(\Omega, \mathfrak{F}, \mathrm{P})$ ($\mathfrak{F}_t \subseteq \mathfrak{F}$; we bear in mind that both the cases of continuous and of discrete time are considered). We use without comments the concepts of Markov time, random function adapted to a given family of σ-algebras, martingale, and local martingale. In the continuous case all martingales and local martingales under consideration will be supposed continuous on the right at every (time) point with probability 1.

We introduce the σ-algebra $\mathcal{P}$ of *predictable* sets as the σ-algebra of subsets of $[t_0, T] \times \Omega$ generated by all sets of the form $(t, T] \times A$, $A \in \mathfrak{F}_t$. A random function is called predictable if it is measurable with respect to $\mathcal{P}$. In the discrete case $\eta(t) = \eta(t, \omega)$ is predictable if and only if for any $t > t_0$ that is a multiple of τ the random variable $\eta(t)$ is measurable with respect to $\mathfrak{F}_{t-\tau}$ and $\eta(t_0)$ is constant. In the case of continuous time, if $\eta(t)$ is taking values in a metric space and is adapted to the family of σ-algebras $\mathfrak{F}_t$, in order that $\eta(t)$ be predictable, left-hand continuity and $\eta(t_0) = \text{const}$ are sufficient (Meyer [1]).

The *compensator* of a real-valued or vector random function $\eta(t)$, $t_0 \le t \le T$, adapted to the given family of σ-algebras (and, in the continuous case, almost surely continuous on the right) is a predictable random function $\tilde{\eta}(t)$, $t_0 \le t \le T$, such that $\tilde{\eta}(t_0) = \eta(t_0)$ almost surely and $\eta(t) - \tilde{\eta}(t)$ is a local martingale; in the continuous case $\tilde{\eta}(t)$ must, with probability 1, also be continuous on the right in t and have bounded variation. If the compensator exists, it is unique with probability 1 (see Meyer [1], p. 297). However, we will not use this uniqueness; nor will we be interested in theorems of existence, under some conditions, of a compensator.

The *bicompensator* of two real-valued random functions $\eta(t)$, $\zeta(t)$, $t_0 \le t \le T$, is defined as the random function $\langle \eta, \zeta \rangle (t) = \widetilde{(\eta - \tilde{\eta})(\zeta - \tilde{\zeta})}(t)$. The *quadratic compensator* of a random function $\eta(t)$ is, in the real case, the random function $\langle \eta, \eta \rangle (t)$, and for a vector random function it is the matrix random function with entries $\langle \eta^i, \eta^j \rangle (t)$. The compensator is an analogue of the expectation, and the quadratic compensator - of the variance. However, this analogy does not extend as far that $\langle \eta, \eta \rangle (t) = \widetilde{\eta^2}(t) - \tilde{\eta}(t)^2$ should hold. Instead

$$\langle \eta, \eta \rangle (t) = \\ = \widetilde{\eta^2}(t) - \tilde{\eta}(t)^2 - 2 \sum_{s = k\tau = t_0}^{t-\tau} (\eta(s) - \tilde{\eta}(s))(\tilde{\eta}(s+\tau) - \tilde{\eta}(s)) \tag{1.2.1}$$

in the discrete case,

$$\langle \eta, \eta \rangle (t) = \widetilde{\eta^2}(t) - \tilde{\eta}(t)^2 - 2 \int_{t_0}^{t} (\eta(s-) - \tilde{\eta}(s-))\, d\tilde{\eta}(s), \tag{1.2.2}$$

$$\langle \eta, \zeta \rangle (t) = \widetilde{\eta\zeta}(t) - \tilde{\eta}(t)\tilde{\zeta}(t) + \\ + \int_{t_0}^{t} (\eta(s-) - \tilde{\eta}(s-))\, d\tilde{\zeta}(s) - \int_{t_0}^{t} (\zeta(s-) - \tilde{\zeta}(s-))\, d\tilde{\eta}(s) \tag{1.2.3}$$

in the continuous case.

1.2.2. Kolmogorov's inequality

If τ_0 is a Markov time taking values in $[t_0, T]$, then for any $\varepsilon > 0$,

$$\mathsf{P}\left\{ \sup_{t_0 \le t \le \tau_0} |\eta(t) - \tilde{\eta}(t)| \ge \varepsilon \right\} \le \varepsilon^{-2} \sum_i \mathsf{M} \langle \eta^i, \eta^i \rangle (\tau_0).$$

1.2.3. Stochastic integrals and their discrete analogues

Here the paralellism of notations in the cases of discrete and continuous time is incomplete. In the discrete case, for random functions $\eta(t) = \eta(t, \omega)$, $t_0 \le t \le T$, and $z(t) = z(t, \omega)$, $t_0 \le t \le T$, adapted to $\{\mathfrak{F}_t\}$ and with values in R^r, we define the random function $\sum_{[t_0, t)} z\,\Delta\eta$ by the formula

$$\sum_{[t_0, t)} z\,\Delta\eta = \sum_{s = k\tau = t_0}^{t-\tau} z(s, \omega)\,[\eta(s+\tau, \omega) - \eta(s, \omega)].$$

One can easily verify that

$$\widetilde{\sum z\,\Delta\eta}\,(t) = \sum_{[t_0, t)} z\,\Delta\tilde{\eta},$$

$$\langle \sum z\,\Delta\eta, \sum z'\,\Delta\eta \rangle = \sum_{[t_0, t)} \sum_{i, j} z_i z'_j\,\Delta\langle \eta^i, \eta^j \rangle$$

(assuming, of course, that the compensator and the quadratic compensator of η exist).

In the continuous case for a right-continuous $\{\mathfrak{F}_t\}$-adapted random function $\eta(t) = \eta(t, \omega)$, $t_0 \le t \le T$, and a predictable random function $z(t) = z(t, \omega)$, $t_0 < t \le T$, the stochastic integral $\int_{t_0}^{t} z(s, \omega)\,d\eta(s)$ is defined if the integrals

$$\int_{t_0}^{t} |z(s, \omega)|\,|d\tilde{\eta}(s)|, \qquad \int_{t_0}^{t} |z(s, \omega)|^2 \sum_i d\langle \eta^i, \eta^i \rangle(s),$$

are finite with probability 1; the stochastic integral is a right-continuous random function adapted to the family of σ-algebras $\{\mathfrak{F}_t\}$, completed with respect to the probability measure and extended by continuity on the right (see Meyer [1]; these conditions are not necessary in order to define the stochastic integral but only sufficient). We have

$$\widetilde{\int z\,d\eta}\,(t) = \int_{t_0}^{t} z(s, \omega)\,d\tilde{\eta}(s),$$

$$\langle \int z\,d\eta, \int z'\,d\eta \rangle(t) = \int_{t_0}^{t} \sum_{i, j} z_i(s, \omega)\,z'_j(s, \omega)\,d\langle \eta^i, \eta^j \rangle(s).$$

A random function admitting a compensator and a quadratic compensator can be represented in form of the sum $\eta(t) = \tilde{\eta}(t) + \eta^c(t) + \eta^d(t)$, where $\eta^c(t)$ is a continuous local martingale and $\eta^d(t)$ is a local martingale that is orthogonal to any continuous local martingale $\zeta(t)$: $\langle\eta^d, \zeta\rangle(t) = 0$ (see Kunita, Watanabe [1], Meyer [1]).

The subsections that follow deal with the continuous case. The corresponding devices can be developed in the discrete case as well, but there is no need for that.

1.2.4. Lévy measures

Let $(Y, \mathcal{Y})$ be a measurable space. We add to Y one element more, denoted by $*$, and in the space $Y \cup \{*\}$ we introduce the σ-algebra $\mathcal{Y}_*$ generated by $\mathcal{Y}$ and $\{*\}$. Let $\eta(t) = \eta(t, \omega)$, $t_0 < t \le T$, be a random function adapted to a given family of σ-algebras $\mathcal{F}_t$, with values in $(Y \cup \{*\}, \mathcal{Y}_*)$, and taking values in Y only for an at most countable set of values of t. By the Lévy measure corresponding to the random function $\eta(t)$ we mean a random function $L(A) = L(A, \omega)$ of sets $A \subseteq (t_0, T] \times Y$ (that are measurable with respect to $\mathcal{B}_{(t_0, T]} \times \mathcal{Y}$) such that

1) for all fixed ω it is a measure (taking possibly infinite values as well) in the argument A;

2) for any non-negative $(\mathcal{P} \times \mathcal{Y})$-measurable function $f(t, \omega, y)$ on $(t_0, T] \times \Omega \times Y$,

a) the random function

$$\int_{(t_0, t]} \int_Y f(s, \omega, y)\, L(ds\, dy)$$

is predictable;

b) if this random function is finite with probability 1, it is the compensator of the random function

$$\zeta(t) = \sum_{\substack{t_0 < s \le t \\ \eta(s) \ne *}} f(s, \omega, \eta(s, \omega)),$$

which is also finite with probability 1;

c) $$\mathsf{M}\zeta(t) = \mathsf{M} \int_{(t_0, t]} \int_Y f(s, \omega, y)\, L(ds\, dy). \tag{1.2.4}$$

It is clear that Condition b) is fullfilled as well for a function f taking values of arbitrary sign if only the integral

$$\int_{(t_0, t]} \int_Y | f(s, \omega, y) | L(ds\, dy)$$

is finite with probability 1.

Let $(\xi(t), \mathsf{P}_{t,x})$, $0 \le t \le T$, be a conservative (non-terminating) time - inhomogeneous Markov process in a metric space X with paths that are right-continuous and have left limits. Let $\lambda_{t,x}(\Gamma)$, $t \in [0, T]$, $x \in X$, $\Gamma \in \mathcal{B}_X$, for fixed t, x be a measure in Γ which is bounded outside arbitrarily small neighbourhoods of the point x; for fixed Γ, let this function be measurable in t, x. Take $\eta(t) = \xi(t)$ if $\xi(t) \neq \xi(t-)$; $\eta(t) = *$ if $\xi(t) = \xi(t-)$. We call $\lambda_{t,x}$ the *Lévy measure, corresponding to the jumps of the process* $(\xi(t), \mathsf{P}_{t,x})$ if for any $t_0 \in [0, T)$ and $x_0 \in X$ the random function

$$L(A, \omega) = \int_{t_0}^{T} \left[\int_X \chi_A(s, y)\, \lambda_{s, \xi(s)}(dy) \right] ds$$

is the Lévy measure for the random function $\eta(t)$, $t_0 < t \le T$, with respect to the probability measure P_{t_0, x_0} and the family of σ-algebras $\mathcal{F}_{t_0 t}$.

Here, for Y we take the state space X itself and for $\eta(t)$ the value of $\xi(t)$ after the jump. Processes with jumps in a linear space can also be described taking $\eta(t)$ to be the size of the jump $\xi(t) - \xi(t-)$.

1.2.5. The compensator for $F(t, \zeta(t))$

Let $\eta(t)$, $t_0 < t \le T$, be a random function with values in $Y \cup \{*\}$, $\mathcal{Y}_*$), let $L(dt\, dy)$ be the associated Lévy measure; let $\zeta(t)$, $t_0 \le t \le T$, be a random function, adapted to the family of σ-algebras $\{\mathcal{F}_t\}$, with values in R^r, and with right-continuous sample functions having left limits; let $\zeta(t_0) = \text{const}$. Let $g(t, \omega, y)$ be a function on $(t_0, T] \times \Omega \times (Y \cup \{*\})$ with values in R^r, measurable with respect to $\mathcal{P} \times \mathcal{Y}_*$, and such that $\zeta(t, \omega) = g(t, \omega, \eta(t, \omega))$, $\zeta(t-, \omega) = g(t, \omega, *)$ (it is clear that the jumps of ζ can only take place at the times when $\eta(t) \neq *$).

Let the random function $\zeta(t)$ admit a compensator and a quadratic compensator, both continuous with respect to t. Let $F(t, x)$ be a real-valued function on $[t_0, T] \times R^r$, once continuously differentiable with respect to the time parameter and twice with respect to the space variables; put $\phi(t) = F(t, \zeta(t))$. Then

$$\tilde{\phi}(t) = F(t_0, \zeta(t_0)) + \int_{t_0}^{t} \frac{\partial F}{\partial s}(s, \zeta(s))\, ds + \int_{t_0}^{t} \sum_i \frac{\partial F}{\partial x^i}(s, \zeta(s))\, d\widetilde{\zeta^i}(s) +$$

$$+ \int_{t_0}^{t} \frac{1}{2} \sum_{i, j} \frac{\partial^2 F}{\partial x^i \partial x^j}(s, \zeta(s))\, d \langle \zeta^{ci}, \zeta^{cj} \rangle (s) +$$

$$+ \int_{(t_0, t]} \int_{Y} [F(s, g(s, \omega, y)) - F(s, \zeta(s)) -$$

$$- \sum_i \frac{\partial F}{\partial x^i}(s, \zeta(s))\, (g^i(s, \omega, y) - \zeta^i(s))]\, L(ds\, dy), \tag{1.2.5}$$

if only

$$\int_{|F(s, g(s, \omega, y))| > C} |F(s, g(s, \omega, y))|\, L(ds\, dy)$$

converges for large C with probability 1.

This is Itô's formula, simplified and adapted to our needs (see Kunita, Watanabe [1], Meyer [1]), from which only the compensator remains and the representation of $\phi(t) - \tilde{\phi}(t)$ in the form of a stochastic integral and a compensated sum over all jumps is left out.

1.3. Compensating operators of Markov processes

1.3.1. Compensating operators

We will consider two classes of conservative (non-terminating) time-inhomogeneous Markov processes $(\xi(t), \mathsf{P}_{t, x})$, $0 \le t \le T$, on a state space $(X, \mathfrak{B})$ (we use the definition of Markov processes in the sense of Dynkin [1]).

The first class is made up of processes with continuous time for which t runs over all real values between 0 and T.

The second class will appear in two closely related forms. The first form is made up by the processes with discrete time running over multiples of some positive τ. The second form is constituted by continuous-time processes changing their position only at times that are multiples of τ and remaining constant on half-intervals $[k\tau, (k+1)\tau)$: $\xi(t) = \xi(k\tau)$ for $k\tau \le t < (k+1)\tau$. We call such processes *τ-processes*.

In the continuous case we say that a measurable real-valued function $f(t, x)$ on $[0, T] \times X$ belongs to the domain of definition $D_{\mathfrak{A}}$ of the *compensating operator* $\mathfrak{A}$ if there exists a measurable function $\phi(t, x)$, $0 \le t \le T$, $x \in X$, such that for any $t_0 \in [0, T]$ and $x_0 \in X$ the compensator of the random function $\eta(t) = f(t, \xi(t))$, $t_0 \le t \le T$, with respect to the probability measure P_{t_0, x_0} and the family of σ-algebras $\{\mathcal{F}_{t_0 t}\}$ is given by the formula

$$\tilde{\eta}(t) = f(t_0, x_0) + \int_{t_0}^{t} \phi(s, \xi(s))\, ds$$

(the integral is assumed to be convergent with probability 1).

If this is true, the value of the compensating operator on the function f is given by $\mathfrak{A}f = \phi$.

We will ignore the questions of uniqueness (almost uniqueness) of a value of the compensating operator, of uniqueness of a process given its compensating operator or some restriction of it, and of construction of a process with given compensating operator.

If a function f depending on the x argument only belongs to the domain of definition of the compensating operator, the *generator* A_t is defined on this function by

$$A_t f(x) = \mathfrak{A}f(t, x).$$

In the case of discrete time the compensating operator $\mathfrak{A}f$ is defined by the equality

$$\widetilde{f(t, \xi(t))} = f(t_0, x_0) + \sum_{s = k\tau = t_0}^{t-\tau} \tau \cdot \mathfrak{A}f(s, \xi(s)).$$

In the discrete case the compensating operator is defined not almost uniquely, but uniquely, namely

$$\mathfrak{A}f(t, x) = \tau^{-1} (\mathsf{M}_{t, x} f(t + \tau, \xi(t + \tau)) - f(t, x)).$$

1.3.2. Locally infinitely divisible processes

A strong Markov process $(\xi(t), \mathsf{P}_{t, x})$ in a $C^{(2)}$-manifold X is called *locally infinitely divisible* if its sample paths are right-continuous and have left limits, the compensating operator is defined for all bounded functions $f(t, x)$ that are once continuously differentiable with respect to t and twice with respect to local coordinates of x, and the value of $\mathfrak{A}f(t, x)$ (of one of the variants of the compensating operator) is given, for any chart (W, ψ), $W \ni x$, by the formula

$$\mathfrak{A}(t, x) = \frac{\partial f}{\partial t}(t, x) + \sum_i b^i(t, x)\frac{\partial f}{\partial x^i}(t, x) +$$

$$+\frac{1}{2}\sum_{i,j} a^{ij}(t, x)\frac{\partial^2 f}{\partial x^i x^j}(t, x) + \int_X [f(t, y) - f(t, x) +$$

$$+\sum_i \frac{\partial f}{\partial x^i}(t, x)(\psi^i(y) - \psi^i(x))]\lambda_{t,x}(dy). \tag{1.3.1}$$

Here the measure $\lambda_{t,x}$ is bounded outside every neighbourhood of the point x, measurable in t, x, and

$$\int_W |\psi(y) - \psi(x)|^2 \lambda_{t,x}(dy) < \infty;$$

ψ is extended outside the region W as a measurable and bounded function; $a^{ij}(t, x)$ and $b^i(t, x)$, for each choice of a chart, depend on t, x in a measurable way; the matrix $(a^{ij}(t, x))$ is symmetric and positive semi-definite, and when we take another chart, this matrix is transformed as a tensor; whereas $b^i(t, x)$ depends both on the choice of a chart and the manner of extending the function ψ beyond the boundary of the region W.

The spirit of this definition is that of "martingale problems" (see Stroock, Varadhan [1], [2]). There are some papers in which infinitely divisible processes are studied (e.g., Grigelionis [1], Komatsu [1], Stroock [1], Lepeltier, Marchal [1]), the equivalence of different definitions is established, and problems of constructing them starting from the local characteristics b^i, a^{ij}, $\lambda_{t,x}$ are considered; nothing of this is touched upon here. The measure $\lambda_{t,x}(dy)$ turns out to be the Lévy measure corresponding to the jumps of the locally infinitely divisible process. The coefficients $a^{ij}(t, x)$ determine the quadratic compensator of the continuous part of the image of the process on the chart. Namely, denote by τ_W the first exit time from W and consider the representation of the random function

$$\eta(t) = \psi(\xi(t \wedge \tau_W)) = \tilde{\eta}(t) + \eta^c(t) + \eta^d(t)$$

in the form of the sum of its compensator with respect to the probability measure corresponding to any initial point, a continuous local martingale, and a local martingale that is orthogonal to all continuous ones. Then

$$\langle \eta^{ci}, \eta^{cj} \rangle(t) = \int_0^{t \wedge \tau_W} a^{ij}(s, \xi(s))\, ds.$$

1.3.3. Lévy measures in the discrete case

In the case of a discrete-time parameter changing with step τ we define the Lévy measure $\lambda_{t,x}$ by

$$\lambda_{t,x}(\Gamma) = \tau^{-1} \mathsf{P}_{t,x}\{\xi(t+\tau) \in \Gamma\}.$$

Here formula (1.3.1) is always valid if we replace $\partial f/\partial t\,(t, x)$ by $\tau^{-1}(f(t+\tau, x) - f(t, x))$ and $f(t, x)$, $\partial f/\partial x^i\,(t, x)$, $f(t, y)$ by $f(t+\tau, x)$, $\partial f/\partial x^i\,(t+\tau, x)$, $f(t+\tau, y)$ and take zero coefficients a^{ij}; the coefficients b^i are expressed by the formula

$$b^i(t, x) = \int_X (\psi^i(y) - \psi^i(x))\, \lambda_{t,x}(dy).$$

1.3.4. Lemmas

We state two auxiliary results, concerning sums over the jumps of a process and the Lévy measures.

Lemma 1.3.1. *Let $V(t, y, x)$, $t \in (t_0, T]$, $y, x \in X$, be a bounded non-negative function that is equal to 0 for $\rho(x, y) < \varepsilon$, $\varepsilon > 0$. Then in the case of a locally infinitely divisible process*

$$\mathsf{M}_{t_0, x_0} \sum_{t_0 < t \le T} V(t, \xi(t-), \xi(t)) =$$

$$= \mathsf{M}_{t_0, x_0} \int_{t_0}^{T} dt \int_X \lambda_{t, \xi(t)}(dx)\, V(t, \xi(t), x); \tag{1.3.2}$$

in the case of a process $(\xi(t), \mathsf{P}_{t,x})$ with $t = k\tau \in [0, T]$:

$$\mathsf{M}_{t_0, x_0} \sum_{t_0 < t = k\tau \le T} V(t, \xi(t-\tau), \xi(t)) =$$

$$= \mathsf{M}_{t_0, x_0} \sum_{t_0 \le t = k\tau < T} \tau \int_X \lambda_{t, \xi(t)}(dx)\, V(t+\tau, \xi(t)\, x). \tag{1.3.3}$$

The sum under the expectation sign in (1.3.2) contains in fact only a finite number of summands since $V(t, \xi(t-), \xi(t))$ is non-zero only in the case of a jump of size $> \varepsilon$.

Proof: Application of formula (1.2.4); in the discrete-time case, a simple calculation. ◊

Lemma 1.3.2. *Let* $V(t_1, y_1, x_1, ..., t_k, y_k, x_k)$, $t_0 < t_1 < ... < t_k \leq T$ $y_i, x_i \in X$, *be a bounded non-negative measurable function vanishing if* $\rho(x_i, y < \varepsilon$ *for some* i ($\varepsilon > 0$). *Then*

$$\mathsf{M}_{t_0, x_0} \sum_{t_0 < t_1 < ... < t_k \leq T} V(t_1, \xi(t_1 -), \xi(t_1), ..., t_k, \xi(t_k -), \xi(t_k)) =$$

$$= \mathsf{M}_{t_0, x_0} \int_{t_0}^{T} dt_1 \int_X \lambda_{t_1, \xi(t_1)}(dx_1) \, \mathsf{M}_{t_1, x_1} \int_{t_1}^{T} dt_2 \int_X \lambda_{t_2, \xi(t_2)}(dx_2) \times$$

$$\times \mathsf{M}_{t_{k-1}, x_{k-1}} \int_{t_{k-1}}^{T} dt_k \int_X \lambda_{t_k, \xi(t_k)}(dx_k) \, V(t_1, \xi(t_1), x_1, ..., t_k, \xi(t_k), x_k). \quad (1.3.$$

Here the last expectation is taken for fixed $\xi(t_1), ..., \xi(t_{k-1})$ *and random* $\xi(t_k$ *i.e. it is to be understood as*

$$\int_\Omega \mathsf{P}_{t_{k-1}, x_{k-1}}(d\omega') \int_{t_{k-1}}^{T} dt_k \int_X \lambda_{t_k, \xi(t_k, \omega')}(dx_k) \times$$

$$\times V(t_1, \xi(t_1, \omega), x_1, ..., t_{k-1}, \xi(t_{k-1}, \omega), x_{k-1}, t_k, \xi(t_k, \omega'), x_k);$$

the last but one, for fixed $\xi(t_1), ..., \xi(t_{k-2})$ *and random* $\xi(t_{k-1})$, *etc.; finally the first one for random* $\xi(t_1)$.

In the discrete case ($t = k\tau$)

$$\mathsf{M}_{t_0, x_0} \sum_{t_0 < t_1 < ... < t_k \leq T} V(t_1, \xi(t_1 - \tau), \xi(t_1), ..., t_k, \xi(t_k - \tau), \xi(t_k)) =$$

$$= \mathsf{M}_{t_0, x_0} \sum_{t_0 \leq t_1 < T} \tau \cdot \int_X \lambda_{t_1, \xi(t_1)}(dx_1) \, \mathsf{M}_{t_1 + \tau, x_1} \sum_{t_1 + \tau \leq t_2 < T} \tau \times ...$$

$$... \times \mathsf{M}_{t_{k-1} + \tau, x_{k-1}} \sum_{t_{k-1} + \tau \leq t_k < T} \tau \cdot \int_X \lambda_{t_k, \xi(t_k)}(dx_k) \times$$

$$\times V(t_1 + \tau, \xi(t_1), x_1 ..., t_k + \tau, \xi(t_k), x_k). \quad (1.3.5)$$

Everywhere in these sums t_i *are multiples of* τ.

The **proof** will be carried out by induction (for $k = 1$ already done). Denote the sum under the expectation sign at the left side of (1.3.4), (1.3.5) by $\sum_{t_0, T}^{k} (V)$. It is sufficient to prove (1.3.4), (1.3.5) for

$$V(t_1, y_1, x_1, \ldots, t_k, y_k, x_k) = V_1(t_1, y_1, x_1) \ldots V_k(t_k, y_k, x_k).$$

In the discrete case the transition from $k-1$ to k is effected using the Markov property:

$$\mathsf{M}_{t_0, x_0} \sum_{t_0, T}^{k} (V_1 \cdot \ldots \cdot V_k) = \sum_{t_0 \le t_1 < T} \mathsf{M}_{t_0, x_0} V_1(t_1 + \tau, \xi(t_1), \xi(t_1 + \tau)) \times$$

$$\times \sum_{t_1 + \tau, T}^{k-1} (V_2 \cdot \ldots \cdot V_k) = \sum_{t_0 \le t_1 < T} \mathsf{M}_{t_0, x_0} V_1(t_1 + \tau, \xi(t_1), \xi(t_1 + \tau)) \times$$

$$\times \mathsf{M}_{t_1 + \tau, \xi(t_1 + \tau)} \sum_{t_1 + \tau, T}^{k-1} (V_2 \cdot \ldots \cdot V_k).$$

In the case of continuous time we introduce the Markov times

$$\tau_0^\varepsilon = t_0, \ \tau_i^\varepsilon = \min\{t > \tau_{i-1}^\varepsilon : \ \rho(\xi(t), \ \xi(t-)) \ge \varepsilon\};$$

if there are no such t we take τ_i^ε and all subsequent τ_{i+1}^ε, τ_{i+2}^ε, ... equal to $+\infty$. It is easy to see that

$$\mathsf{M}_{t_0, x_0} \sum_{t_0, T}^{k} (V_1 \cdot \ldots \cdot V_k) = \mathsf{M}_{t_0, x_0} \sum_{i=1}^{\infty} V_1(\tau_i^\varepsilon, \xi(\tau_i^\varepsilon -), \xi(\tau_i^\varepsilon)) \sum_{\tau_i^\varepsilon, T}^{k-1} (V_2 \cdot \ldots \cdot V_k).$$

Here we have assumed $V_1(+\infty, y, x) = 0$. Invert the order of taking the expectation and the sum and make use of the strong Markov property with respect to the time τ_i^ε in the i-th term of the sum. The random variable $V_1(\tau_i^\varepsilon, \xi(\tau_i^\varepsilon -), \xi(\tau_i^\varepsilon))$ can be taken outside the conditional expectation sign and under it remains $G(\tau_i^\varepsilon, \xi(\tau_i^\varepsilon))$, where

$$G(t, x) = \mathsf{M}_{t, x} \sum_{t, T}^{k-1} (V_2 \cdot \ldots \cdot V_k).$$

By the previous lemma we find that

$$\mathsf{M}_{t_0, x_0} \sum_{t_0, T}^{k} (V_1 \cdot \ldots \cdot V_k) =$$

$$= \sum_{i=1}^{\infty} \mathsf{M}_{t_0, x_0} V_1(\tau_i^\varepsilon, \xi(\tau_i^\varepsilon -), \xi(\tau_i^\varepsilon))\, G(\tau_i^\varepsilon, \xi(\tau_i^\varepsilon)) =$$

$$= \mathsf{M}_{t_0, x_0} \sum_{i=1}^{\infty} V_1 (\tau_i^\varepsilon, \xi (\tau_i^\varepsilon -), \xi (\tau_i^\varepsilon)) \, G (\tau_i^\varepsilon, \xi (\tau_i^\varepsilon)) =$$

$$= \mathsf{M}_{t_0, x_0} \sum_{t_0 < t \le T} V_1 (t, \xi (t -), \xi (t)) \, G (t, \xi (t)) =$$

$$= \mathsf{M}_{t_0, x_0} \int_{t_0}^{T} dt \int_X \lambda_{t, \xi (t)} (dx) \, V_1 (t, \xi (t), x) \, G (t, x).$$

Substituting the expression for $G (t, x)$, which holds by virtue of the induction assumption, we obtain (1.3.4). ◊

CHAPTER 2

ESTIMATES ASSOCIATED WITH THE ACTION FUNCTIONAL FOR MARKOV PROCESSES

2.1. The action functional

2.1.2. The cumulant

In § 2.1 - § 2.3 we will consider Markov processes in R^r satisfying an analogue of Cramér's condition of finiteness of exponential moments. We will consider two classes of such processes: those with a discrete parameter whose values are multiples of some $\tau > 0$, and locally infinitely divisible ones with compensating operator given by the formula

$$\mathfrak{A}f(t, x) = \frac{\partial f}{\partial t}(t, x) + \sum_i b^i(t, x) \frac{\partial f}{\partial x^i}(t, x) + \frac{1}{2} \sum_{i, j} a^{ij}(t, x) \frac{\partial^2 f}{\partial x^i x^j}(t, x) +$$

$$+ \int_{R^r} [f(t, y) - f(t, x) - \sum_i \frac{\partial f}{\partial x^i}(t, x)(y^i - x^i)] \lambda_{t, x}(dy). \qquad (2.1.1)$$

In order that the integral converges for bounded smooth f, it is necessary to impose on the measure $\lambda_{t, x}$ some conditions at infinity, so that this form of the operator is less general than (1.2.5) where the function ψ is extended outside a neighbourhood of x as a bounded function; but anyway we are going to impose even more restrictions on the coefficients of the operator and the measure $\lambda_{t, x}$.

The *cumulant* of the process $(\xi(t), \mathsf{P}_{t, x})$ is a function $G(t, x; z)$ of $t \in [0, T]$, $x \in R^r$ and $z \in R^r$, taking values in $(-\infty, \infty]$. In the discrete case it is defined by the formula

$$G(t, x; z) = \tau^{-1} \ln \mathsf{M}_{t, x} \exp\{z(\xi(t + \tau) - x)\}. \qquad (2.1.2)$$

(In the papers Wentzell [7] the cumulant was defined in the discrete case somewhat differently: without the factor τ^{-1}; we introduce this factor in order to obtain greater parallelism in considering discrete time and continuous time.) In the continuous case

$$G(t, x; z) = \sum_i z_i b^i(t, x) + \frac{1}{2} \sum_{i, j} z_i z_j a^{ij}(t, x) +$$

$$+ \int_{R^r} [e^{z(y - x)} - 1 - z(y - x)] \lambda_{t, x}(dy). \qquad (2.1.3)$$

The term *cumulant* is used in a similar way in Gikhman, Skorokhod [1].

It is clear that the function $G(t, x; z)$ is measurable with respect to t, x, z and $G(t, x; 0) \equiv 0$. We can easily verify that for fixed t, x this function is downward convex and lower semi-continuous with respect to z; in the interior of its domain of finiteness it is analytic with respect to z and its matrix of second-order derivatives is strictly positive definite.

We yet impose a restriction on G:

A. $G(t, x; z) \le \overline{G}(z)$ for all t, x, z, where $\overline{G}(z)$ is a non-negative downward convex function, $\overline{G}(0) = 0$, $\overline{G}(z) < \infty$ for z in some open set Z containing the point 0.

We can always find a function $\overline{G}$ satisfying these conditions except the condition of boundedness in a neighbourhood of zero, by taking $\overline{G}(z) = 0 \vee \sup_{t,x} G(t, x; z)$; so **A** is a boundedness condition.

If Condition **A** is fulfilled, the function $G(t, x; z)$ is bounded from above and from below for all t, x, and all z changing over any compact subset of Z.

2.1.2. Expression for the compensator and quadratic compensator in terms of the cumulant

It is easily seen that under Condition **A** the function G is infinitely differentiable in z at the point $z \equiv 0$ and that its derivatives are related to the moments. The compensator and the quadratic compensator of $\xi(t)$, $t \in [t_0, T]$, with respect to all probability measures P_{t_0, x_0}, are expressed in the following way:

in the discrete case:

$$\tilde{\xi}(t) = x_0 + \sum_{s = k\tau = t_0}^{t-\tau} \nabla_z G(s, \xi(s); 0) \cdot \tau,$$

$$\langle \xi^i, \xi^j \rangle (t) = \sum_{s = k\tau = t_0}^{t-\tau} \frac{\partial^2 G}{\partial z_i \partial z_j}(s, \xi(s); 0) \cdot \tau;$$

while in the continuous case the sums are replaced by the integrals from t_0 to t. In the continuous case this is deduced from the fact that

$$\tilde{\xi}(t) = x_0 + \int_{t_0}^{t} b(s, \xi(s))\, ds, \quad \langle \xi^i, \xi^j \rangle (t) = \int_{t_0}^{t} A^{ij}(s, \xi(s))\, ds,$$

where

$$b = (b^1, \dots, b^r), \quad A^{ij}(s, x) = a^{ij}(s, x) + \int_{R^r} (y^i - x^i)(y^j - x^j)\, \lambda_{t, x}(dy).$$

The functions $b(s, x)$ and $(A^{ij}(s, x))$ are bounded uniformly with respect to s, x, because for sufficiently small positive ε (small such that $\{z : |z| \le \varepsilon\} \subset Z$) we have

$$|b(s, x)| = |\nabla_z G(s, x; 0)| \le \varepsilon^{-1} \sup_{|z| \le \varepsilon} \overline{G}(z),$$

$$|A^{ij}(s, x)| \le (A^{ii}(s, x) + A^{jj}(s, x)) / 2,$$

$$A^{ii}(s, x) = a^{ii}(s, x) + \int_{R^r} (y^i - x^i)\, \lambda_{s, x}(dy) \le$$

$$\le a^{ii}(s, x) + \varepsilon^{-2} \int_{R^r} [e^{\varepsilon (y^i - x^i)} - 1 + e^{-\varepsilon (y^i - x^i)} - 1]\, \lambda_{s, x}(dy) =$$

$$= \varepsilon^{-2} [G(s, x; 0, \dots, \underset{i}{\varepsilon}, \dots, 0) + G(s, x; 0, \dots, -\underset{i}{\varepsilon}, \dots, 0)] \le$$

$$\le 2\varepsilon^{-2} \sup_{|z| \le \varepsilon} \overline{G}(z).$$

2.1.3. The functional $\pi(t_0, t)$

Now we state the main property of the cumulant used by us.

The case of discrete time:

Lemma 2.1.1. *Let t_0 be a multiple of τ between 0 and T, and $z(t, \omega)$, $t \in [t_0, T] \in [t_0, T]$, a random function adapted to the family of σ-algebras $\{\mathfrak{F}_{[t_0, t]}\}$ with values in Z. Then the random function*

$$\pi(t_0, t) =$$

$$= \exp\left\{ \sum_{s = k\tau = t_0}^{t - \tau} z(s, \omega)(\xi(s + \tau) - \xi(s)) - \sum_{s = k\tau = t_0}^{T - \tau} G(s, \xi(s); z(s, \omega)) \cdot \tau \right\},$$

$$t \in [t_0, T],$$

is a martingale with respect to any of the probability measures P_{t_0, x_0} and the family of σ-algebras $\{\mathfrak{F}_{[t_0, t]}\}$.

Proof. Applying the Markov property we verify that

$$\mathsf{M}_{t_0, x_0}(\pi(t_0, t + \tau)\, \pi(t_0, t)^{-1} \mid \mathfrak{F}_{[t_0, t]}) = 1. \ \Diamond$$

The case of continuous time:

Lemma 2.1.1'. *Let $t_0 \in [0, T]$; let $z(t, \omega)$, $t_0 < t \le T$, be a predictable random function all values of which belong to some compact subset $K \subset Z$. Then the random function*

$$\pi(t_0, t) = \exp\left\{ \int_{t_0}^{t} z(s, \omega)\, d\xi(s) - \int_{t_0}^{t} G(s, \xi(s); z(s, \omega))\, ds \right\}, \quad t \in [t_0, T],$$

is a martingale with respect to any one of the probability measures P_{t_0, x_0} and the family of σ-algebras $\mathfrak{F}_{t_0 t}$.

Here we consider the completed σ-algebras $\mathfrak{F}_{t_0 t}$ because of the use of stochastic integrals.

Proof. Using formula (1.2.4) we verify that $\tilde{\pi}(t_0, t) = 1$, so that $\pi(t_0, t)$ is a local martingale. Since $\pi(t_0, t) > 0$, it follows that $\mathsf{M}_{t_0, x_0} \pi(t_0, \tau_0) \le 1$ for any Markov time τ_0, $t_0 \le \tau_0 \le T$. In order to prove that $\pi(t_0, t)$ is a martingale it is sufficient to establish uniform integrability of $\pi(t_0, \tau_0)$ for all Markov times τ_0, $t_0 \le \tau_0 \le T$; and for this it is sufficient to verify the boundedness of $\mathsf{M}_{t_0, x_0} \pi(t_0, \tau_0)^{1+\varepsilon}$ for some $\varepsilon > 0$. But for sufficiently small ε the compact set $(1 + \varepsilon) K \subset Z$; so

$$\pi'(t_0, t) = \exp\left\{ \int_{t_0}^{t} (1 + \varepsilon) z(s, \omega)\, d\xi(s) - \int_{t_0}^{t} G(s, \xi(s); (1 + \varepsilon) z(s, \omega))\, ds \right\}$$

is a local martingale, $\mathsf{M}_{t_0, x_0} \pi'(t_0, \tau_0) \le 1$, and

$$\mathsf{M}_{t_0, x_0} \pi(t_0, \tau_0)^{1+\varepsilon} \le$$

$$\le \exp\{(T - t_0)[\sup\{G(t, x; (1 + \varepsilon) z) : t \in [t_0, T), x \in R^r, z \in K\} +$$

$$+ (1 + \varepsilon) \inf\{G(t, x; z) : t \in [t_0, T), x \in R^r, z \in K\}]\}. \ \Diamond$$

2.1.4. The functions $H(t, x; u)$, $\underline{H}(u)$

Put $H(t, x; u) \leftrightarrow G(t, x; z)$; the function H is non-negative, because $G(t, x; 0) = 0$, and measurable with respect to t, x, u. The set on which this function is finite is related to the possible sizes of jumps of the random function $\xi(t)$.

Examples. For discrete time: if the distribution of $\tau^{-1}[\xi(t+\tau)-\xi(t)]$ (for the process starting from the point x in the plane at time t) is the mixture of a distribution with positive density in some disk and the distribution concentrated at some point u_0 of its periphery, then $G(t, x; z)$ is finite for all z; $H(t, x; u) = \infty$ outside this disk and at the periphery, except at the point u_0; inside the disk and at u_0 the function H takes finite values, and it tends to ∞ when u approaches from the inside a point on the periphery not equal to u_0. In the continuous case, if x is an interior point of the convex hull of the support of the measure $\lambda_{t,x}$, or if the matrix $(a^{ij}(t, x))$ is strictly positive definite, then $G(t, x; z)$ tends to infinity faster than linearly in every direction, and $H(t, x; u)$ is finite everywhere. If $a^{ij}(t, x) = 0$ and the measure $\lambda_{t,x}$ is concentrated at two points visible from x at some angle, then $H(t, x; u)$ is finite at all points of this angle (including the boundary), shifted by a certain vector. If

$$a^{ij}(t, x) = 0,\ \lambda_{t,x}(R^r \setminus \{x\}) < \infty$$

and $\lambda_{t,x}$ has a positive density on a convex open set that does not contain x, then $H(t, x; u) < \infty$ inside the angle at which this set is seen from x, shifted by some vector, and at the vertex, but not on the remainder of the boundary. If $\lambda_{t,x}(R^r \setminus \{x\}) = \infty$ (which can hold only if x belongs to the boundary of the above-mentioned convex set) but $\int (y-x)\,\lambda_{t,x}(dy)$ converges, the function H is infinite at the vertex too. However, if this integral is divergent in some of the interior directions of the angle, then $H(t, x; u)$ is finite everywhere.

Let us introduce the function $\underline{H}(u) \leftrightarrow \overline{G}(z)$. It is easily seen that $0 \le \underline{H}(u) \le H(t, x; u)$ for all t, x, u. Denote by $\overline{U}$ the closure of the set $\{u: \underline{H}(u) < \infty\}$. The set $\overline{U}$ is necessarily convex.

Lemma 2.1.2. *Let z be a non-zero vector, let d be a real number and let $0 \le t \le t' \le T$. If $\underline{H}(u) = \infty$ for all u such that $zu > d$, then almost surely $z(\xi(t') - \xi(t)) \le (t'-t)\,d$.*

Proof. For a natural number n and $\varepsilon > 0$ use the exponential Chebyshev inequality:

$$\begin{aligned} &\mathsf{P}_{t,x}\{z(\xi(t')-\xi(t)) > (t'-t)(d+\varepsilon)\} \le \\ &\le \mathsf{M}_{t,x}\exp\{nz(\xi(t')-\xi(t))\}/\exp\{n(t'-t)(d+\varepsilon)\} \le \\ &\le \exp\{(t'-t)[\overline{G}(nz)-nd] - n(t'-t)\varepsilon\}. \end{aligned} \tag{2.1.4}$$

But $\overline{G}(nz) = \sup_u [nzu - \underline{H}(u)]$; moreover, the least upper bound can be taken over those u for which $\underline{H}(u) < \infty$. The scalar product here does not exceed nd, and $\underline{H}(u) \geq 0$; so the expression (2.1.4) does not exceed $\exp\{-n(t'-t)\varepsilon\}$. Putting $n \to \infty$ we find that the probability is equal to 0; to conclude the proof we let ε tend to 0. ◊

Corollary. For any $0 \leq t < t' \leq T$, almost surely $\xi(t') - \xi(t) \in (t'-t) \cdot \overline{U}$. For the proof we represent $\overline{U}$ as a countable intersection of half-spaces of the form $\{u: zu \leq d\}$.

2.1.5. The action functional; the set $\Phi_{x\,[T_1, T_2]}(i)$

Define the *action functional* $I_{T_1, T_2}(\phi)$ for the process $\xi(t)$ on a time interval $[T_1, T_2] \subseteq [0, T]$ as follows on functions $\phi(t)$, $t \in [T_1, T_2]$, taking values in R^r. In the discrete case:

$$I_{T_1, T_2}(\phi) = \sum_{s = k\tau = t_0}^{t-\tau} H\left(s, \phi(s); \frac{\phi(s+\tau) - \phi(s)}{\tau}\right) \cdot \tau;$$

and in the continuous case for absolutely continuous ϕ:

$$I_{T_1, T_2}(\phi) = \int_{T_1}^{T_2} H(s, \phi(s); \dot{\phi}(s))\, ds,$$

while for other ϕ we put $I_{T_1, T_2}(\phi) = +\infty$.

For $x \in R^r$, $[T_1, T_2] \subseteq [0, T]$, $i \geq 0$ we consider the set $\Phi_{x;\,[T_1, T_2]}(i) = \{\phi: \phi(T_1) = x,\ I_{T_1, T_2}(\phi) \leq i\}$.

2.2. Derivation of the lower estimate for the probability of passing through a tube

2.2.1. Theorem 2.2.1. *Let $(\xi(t), \mathsf{P}_{t,x})$ be a Markov process belonging to one of the two classes introduced in the previous section; let the corresponding cumulant satisfy Condition* **A**. *Let $\phi(t)$, $0 \leq t \leq T$, be a function with values in R^r, absolutely continuous in the continuous case. In the discrete case, let the function $H(t, x; u)$ be differentiable with respect to u at the point $(\phi(t+\tau) - \phi(t))/\tau$ for all x and all $t \in [0, T)$ that are multiples of τ, and let all values of the function $z(t, x) = \nabla_u H(t, x; (\phi(t+\tau) - \phi(t))/\tau)$ belong to a compact subset of Z; in the continuous case, let the same be true for all x and almost all $t \in [0, T)$ for the values* *of* $z(t, x) = \nabla_u H(t, x; \dot{\phi}(t))$.

Take, in the discrete case, for $0 \le t < T$,

$$D(t) = \sup_x \sum_i \frac{\partial^2 G}{\partial z_i^2}(t, x; z(t, x)), \tag{2.2.1}$$

$$ZDZ(t) = \sup_x \sum_{i,j} \frac{\partial^2 G}{\partial z_i \partial z_j}(t, x; z(t, x))\, z_i(t, x)\, z_j(t, x), \tag{2.2.2}$$

$$\delta' = 2\left(\sum_{t = k\tau = 0}^{T-\tau} D(t)\cdot\tau\right)^{1/2}, \tag{2.2.3}$$

$\Delta H(t) =$

$$= \sup_{|x - \phi(t)| < \delta'}\left[H\left(t, x; \frac{\phi(t+\tau) - \phi(t)}{\tau}\right) - H\left(t, \phi(t); \frac{\phi(t+\tau) - \phi(t)}{\tau}\right)\right]; \tag{2.2.4}$$

and in the continuous case, the same, but with $(\phi(t+\tau) - \phi(t))/\tau$ *instead of* $\dot{\phi}(t)$, *and instead of the sum in* (2.2.3) *the integral*:

$$\delta' = 2\left(\int_0^T D(t)\,dt\right)^{1/2}. \tag{2.2.3'}$$

Then, if $\delta \ge \delta'$ *and* $x_0 = \phi(0)$, *in the discrete case*

$$\mathsf{P}_{0, x_0}\{\rho_{0,T}(\xi, \phi) < \delta\} \ge$$

$$\ge \frac{1}{2}\exp\left\{-I_{0,T}(\phi) - 2\left(\sum_{t = k\tau = 0}^{T-\tau} ZDZ(t)\cdot\tau\right)^{1/2} - \sum_{\tau = \kappa\tau = 0}^{T-\tau} \Delta H(t)\cdot\tau\right\}; \tag{2.2.5}$$

and in the continuous case

$$\mathsf{P}_{0, x_0}\{\rho_{0,T}(\xi, \phi) < \delta\} \ge$$

$$\ge \frac{1}{2}\exp\left\{-I_{0,T}(\phi) - 2\left(\int_0^T ZDZ(t)\,dt\right)^{1/2} - \int_0^T \Delta H(t)\,dt\right\}. \tag{2.2.5'}$$

Proof. Introduce the *generalized Cramér's transformation* corresponding to an arbitrary measurable function $z(t, x)$ taking values in a compact subset of Z (first introduced in Wentzell [4], [5]; for the source of the idea see Cramér [1]). In the discrete case take

$$\pi(0, t) = \\ = \exp\left\{ \sum_{s = k\tau = 0}^{t-\tau} z(s, \xi(s))(\xi(s+\tau) - \xi(s)) - \sum_{s = k\tau = 0}^{t-\tau} G(s, \xi(s); z(s, \xi(s)) \cdot \tau \right\};$$

and in the continuous case take

$$\pi(0, t) = \exp\left\{ \int_0^t z(s, \xi(s-))\, d\xi(s) - \int_0^t G(s, \xi(s); z(s, \xi(s)))\, ds \right\}$$

(bear in mind that the random function $\xi(s-)$ is predictable).

The generalized Cramér's transformation corresponding to $z(t, x)$ is the transformation of replacing the probability measure P_{0, x_0} with the measure P^z_{0, x_0} defined, for events $A \in \mathfrak{F}_{0T}$, by the formula

$$\mathsf{P}^z_{0, x_0}(A) = \mathsf{M}_{0, x_0}(A; \pi(0, T)).$$

According to Lemmas 2.1.1, 2.1.1', P^z_{0, x_0} is a probability measure.

2.2.2. Characteristics of the transformed process

The compensator and the quadratic compensator $\xi(t)$ with respect to this probability measure are given by the following formulas: in the discrete case

$$\tilde{\xi}^z(t) = x_0 + \sum_{s = k\tau = 0}^{t-\tau} \nabla_z G(s, \xi(s); \ z(s, \xi(s))) \cdot \tau, \tag{2.2.6}$$

$$\langle \xi^i, \xi^j \rangle^z(t) = \sum_{s = k\tau = 0}^{t-\tau} \frac{\partial^2 G}{\partial z_i \partial z_j}(s, \xi(s); \ z(s, \xi(s))) \cdot \tau; \tag{2.2.7}$$

in the continuous case

$$\tilde{\xi}^z(t) = x_0 + \int_0^t \nabla_z G(s, \xi(s); \ z(s, \xi(s)))\, ds, \tag{2.2.8}$$

$$\langle \xi^i, \xi^j \rangle^z(t) = \int_0^t \frac{\partial^2 G}{\partial z_i \partial z_j}(s, \xi(s); \ z(s, \xi(s)))\, ds. \tag{2.2.9}$$

In order to prove that formula (2.2.6) (or (2.2.8)) does give the compensator of $\xi(t)$ it is sufficient to verify that there exists a sequence of Markov times $T_n \uparrow T$ such that for any $t_1 \le t$ belonging to $[0, T]$ and any event $A \in \mathfrak{F}_{0t_1}$, the expectation

$$\mathsf{M}^z_{0,\,x_0}(A;\ \xi(t\wedge T_n)-\tilde{\xi}^z(t\wedge T_n)-\xi(t_1\wedge T_n)+\tilde{\xi}^z(t_1\vee T_n))=$$

$$=\mathsf{M}_{0,\,x_0}(A;\ \pi(0,T)\,[\xi(t\wedge T_n)-\tilde{\xi}^z(t\wedge T_n)-$$

$$-\xi(t_1\wedge T_n)+\tilde{\xi}^z(t_1\wedge T_n)]) \tag{2.2.10}$$

vanishes. Making use of the fact that $\pi(0, t)$ is a martingale we reduce the expectation (2.2.10) to

$$\mathsf{M}^z_{0,\,x_0}(A;\ \pi(0,t\wedge T_n)\,[\xi(t\wedge T_n)-\tilde{\xi}^z(t\wedge T_n)-\xi(t_1\wedge T_n)+\tilde{\xi}^z(t_1\wedge T_n)]).$$

So we have to verify that the compensator of $\pi(0, t)\,[\xi(t) - \tilde{\xi}^z(t)]$ with respect to the initial probability measure is equal to 0.

In the discrete case

$$\mathsf{M}_{0,\,x_0}\{\pi(0,t+\tau)\,[\xi(t+\tau)-\tilde{\xi}^z(t+\tau)]\mid\mathfrak{F}_{0t}\}=$$
$$=\pi(0,t)\,\mathsf{M}_{0,\,x_0}\{\xi(t+\tau)\exp\{z(t,\xi(t))(\xi(t+\tau)-\xi(t))-$$
$$-\tau G(t,\xi(t);z(t,\xi(t)))\}\mid\mathfrak{F}_{0t}\}-\tilde{\xi}^z(t+\tau)\,\mathsf{M}_{0,\,x_0}\{\pi(0,t+\tau)\mid\mathfrak{F}_{0t}\}=$$
$$=\pi(0,t)\exp\{-\tau G(t,\xi(t);z(t,\xi(t)))\}\times$$
$$\times\,\mathsf{M}_{t,\,x}\,\xi(t+\tau)\exp\{z(t,x)(\xi(t+\tau)-x)\}\mid_{x=\xi(t)}-$$
$$-\pi(0,t)\,\tilde{\xi}^z(t+\tau). \tag{2.2.11}$$

Here we have used the Markov property with respect to the (time) point t. But

$$\mathsf{M}_{t,\,x}\,\xi(t+\tau)\exp\{z(t,x)(\xi(t+\tau)-x)\}=$$
$$=x\,\mathsf{M}_{t,\,x}\exp\{z(t,x)(\xi(t+\tau)-x)\}+$$
$$+\nabla_z\,\mathsf{M}_{t,\,x}\exp\{z(\xi(t+\tau)-x)\}\mid_{z=z(t,x)}=$$
$$=x\exp\{\tau\cdot G(t,x;z(t,x))\}+$$
$$+\exp\{\tau\cdot G(t,x;z(t,x))\}\cdot\tau\cdot\nabla_z G(t,x;z(t,x)).$$

The conditional expectation (2.2.11) can be rewritten in the form

$$\pi(0,t)\,[\xi(t)+\tau\cdot\nabla_z G(t,x;z(t,x))-\tilde{\xi}^z(t+\tau)]=\pi(0,t)\,[\xi(t)-\tilde{\xi}^z(t)],$$

which was to be proved.

The formula for the quadratic compensator in the discrete case is verified in a similar way.

Now we verify formulas (2.2.8), (2.2.9) for the compensator and the quadratic compensator of $\xi(t)$ with respect to the transformed probability in the continuous case. To this end we choose ε in such a way that the closed ε-neighbourhood of the compactum containing all values of $z(t, x)$ is still contained in Z, and prove that, for a function $f(t, x)$ growing not faster than $e^{\varepsilon |x|}$ and once continuously differentiable with respect to t and twice with respect to x, the random function

$$\pi(0, t)\,[f(t, \xi(t)) - f(0, x_0) - \int_0^t \mathfrak{A}^z f(s, \xi(s))\, ds] \tag{2.2.12}$$

is a local martingale with respect to P_{0, x_0}, where

$$\mathfrak{A}^z f(t, x) = \frac{\partial f}{\partial t}(t, x) + \sum_i \frac{\partial f}{\partial x^i}(t, x)\,\frac{\partial G}{\partial z_i}(t, x; z(t, x)) + \frac{1}{2}\sum_{i,j} a^{ij}(t, x)\,\frac{\partial^2 f}{\partial x^i \partial x^j}(t, x) +$$

$$+ \int_{R^r} [f(t, y) - f(t, x) - \sum_i \frac{\partial f}{\partial x^i}(t, x)\,(y^i - x^i)]\; e^{z(t, x)(y - x)}\, \lambda_{t, x}(dy).$$

To prove this we put $\psi(t) = \pi(0, t) f(t, \xi(t))$ and evaluate the compensator $\tilde{\psi}(t)$ by means of formula (1.2.5). For $\eta(t)$ we take the random function equal to $\xi(t)$ at its discontinuity points and to $*$ at the points of continuity.

Take

$$\xi^0(t) = \int_0^t z(s, \xi(s-))\, d\xi(s) - \int_0^t G(s, \xi(s);\; z(s, \xi(s)))\, ds.$$

We have

$$\psi(t) = F(t, \xi^0(t), \xi^1(t), \ldots, \xi^r(t)),$$

where

$$F(t, x^0, x^1, \ldots, x^r) = e^{x^0} f(t, x^1, \ldots, x^r).$$

We list the corresponding functions g^k, the compensators of $\xi^k(t)$ and the bicompensators of their continuous parts:

$$g^0(t, \omega, y) = \xi^0(t-) + z(t, \xi(t-))\,(y - \xi(t-)),$$

$$g^i(t, \omega, y) = y^i, \quad 1 \le i \le r,$$

$$g^k(t, \omega, *) = \xi^k(t-), \quad 0 \le k \le r;$$

$$\tilde{\xi}^0(t) = \int_0^t [z(s, \xi(s))\, b(s, \xi(s)) - G(s, \xi(s);\; z(s, \xi(s)))]\, ds,$$

$$\tilde{\xi}^i(t) = x_0^i + \int_0^t b^i(s, \xi(s))\, ds, \quad 1 \le i \le r,$$

$$\langle \xi^{c0}, \xi^{c0} \rangle (t) = \int_0^t \sum_{i,j} a^{ij}(s, \xi(s))\, z_i(s, \xi(s))\, z_j(s, \xi(s))\, ds,$$

$$\langle \xi^{c0}, \xi^{ci} \rangle (t) = \int_0^t \sum_{j} a^{ij}(s, \xi(s))\, z_j(s, \xi(s))\, ds,$$

$$\langle \xi^{ci}, \xi^{cj} \rangle (t) = \int_0^t a^{ij}(s, \xi(s))\, ds, \quad 1 \le i \le r.$$

Applying (1.2.5) yields:

$$\tilde{\psi}(t) = f(0, x_0) + \int_0^t e^{\xi^0(s)} \Bigg\{ \frac{\partial f}{\partial s}(s, \xi(s)) +$$

$$+ f(s, \xi(s))\, [z(s, \xi(s))\, b(s, \xi(s)) - G(s, \xi(s);\ z(s, \xi(s)))] +$$

$$+ \sum_{i=1}^{r} \frac{\partial f}{\partial x^i}(s, \xi(s))\, b^i(s, \xi(s)) +$$

$$+ \frac{1}{2} f(s, \xi(s)) \sum_{i,j=1}^{r} a^{ij}(s, \xi(s))\, z_i(s, \xi(s))\, z_j(s, \xi(s)) +$$

$$+ \sum_{i,j=1}^{r} \frac{\partial f}{\partial x^i}(s, \xi(s))\, a^{ij}(s, \xi(s))\, z_j(s, \xi(s)) +$$

$$+ \frac{1}{2} \sum_{i,j=1}^{r} \frac{\partial^2 f}{\partial x^i \partial x^j}(s, \xi(s))\, a^{ij}(s, \xi(s)) + \int_{R^r} [e^{z(s, \xi(s))(y - \xi(s))} f(s, y) -$$

$$- f(s, \xi(s)) - f(s, \xi(s)) \cdot z(s, \xi(s))\, (y - \xi(s)) -$$

$$- \sum_{i=1}^{r} \frac{\partial f}{\partial x^i}(s, \xi(s))\, (y^i - \xi^i(s))]\, \lambda_{s, \xi(s)}(dy) \Bigg\}\, ds.$$

Here we changed $\xi(s-)$ to $\xi(s)$, because the set of discontinuity points has Lebesgue measure zero. The last integral between brackets can be rewritten as

$$\int_{R^r} [f(s, y) - f(s, \xi(s)) - \sum_{i=1}^{r} \frac{\partial f}{\partial x^i}(s, \xi(s))\, (y^i - \xi^i(s))] \times$$

$$\times e^{z(s, \xi(s))(y - \xi(s))}\, \lambda_{s, \xi(s)}(dy) +$$

$$+ f(s, \xi(s)) \cdot \int_{R^r} [e^{z(s, \xi(s))(y - \xi(s))} - 1 - z(s, \xi(s))(y - \xi(s))] \lambda_{s, \xi(s)}(dy) +$$

$$+ \sum_{i=1}^{r} \frac{\partial f}{\partial x^i}(s, \xi(s)) \cdot \int_{R^r} (y^i - \xi^i(s)) [e^{z(s, \xi(s))(y - \xi(s))} - 1] \lambda_{s, \xi(s)}(dy).$$

The coefficients at $f(s, \xi(s))$ and at $\frac{\partial f}{\partial x^i}(s, \xi(s))$ are equal to

$$G(s, \xi(s); z(s, \xi(s))) - z(s, \xi(s)) b(s, \xi(s)) -$$

$$- \frac{1}{2} \sum_{i, j = 1}^{r} a^{ij}(s, \xi(s)) z_i(s, \xi(s)) z_j(s, \xi(s))$$

and

$$\frac{\partial G}{\partial x^i}(s, \xi(s); z(s, \xi(s))) - b^i(s, \xi(s)) - \sum_{j=1}^{r} a^{ij}(s, \xi(s)) z_j(s, \xi(s)),$$

respectively. Gathering similar terms, we obtain

$$\tilde{\psi}(t) = f(0, x_0) + \int_0^t \pi(0, s) \mathfrak{A}^z f(s, \xi(s)) \, ds.$$

The compensator of $\pi(0, t) f(0, x_0)$ is equal to $f(0, x_0)$; formula (2.2.12) still includes the part containing the integral. Denote this integral by $\tilde{\eta}^z(t)$. Then we have

$$\pi(0, t) \tilde{\eta}^z(t) = \int_0^t \tilde{\eta}^z(s) \, d\pi(0, s) + \int_0^t \pi(0, s) \, d\tilde{\eta}^z(s).$$

The compensator of this expression is

$$\int_0^t \tilde{\eta}^z(s) \, d\tilde{\pi}(0, s) + \int_0^t \pi(0, s) \, d\tilde{\eta}^z(s) = \int_0^t \pi(0, s) \mathfrak{A}^z f(s, \xi(s)) \, ds.$$

Finally we find that the compensator of the expression (2.2.12) is 0, which proves our statement. Now we obtain the expression (2.2.8) for the compensator of $\xi^i(t)$ with respect to the probability P^z_{0, x_0}, taking $f(t, x) = x^i$, and (2.2.9), taking $f(t, x) = x^i x^j$ and making use of formula (1.2.3).

Note that if we introduce $\pi(t_0, t)$ in the obvious way, and for $t_0 \in [0, T]$, $x \in R^r$, define the measures $\mathsf{P}^z_{t_0, x}$ on $\mathfrak{F}_{t_0 T}$ by $\mathsf{P}^z_{t_0, x}(A) = \mathsf{M}_{t_0, x}(A; \pi(t_0, T))$, then $(\xi(t), \mathsf{P}^z_{t, x})$, $0 \le t \le T$, turns out to be a Markov process again (in the continuous case, a locally infinitely divisible one) but with different characteristics. In this way we could have deduced the required results from those by Grigelionis [1], Komatsu [1] concerning densities in the function space of distributions of a locally infinitely divisible process with respect to another. We do not proceed in such a way, first, because these results are based upon the generalized Itô formula also; and, secondly, in order to be able to use the same method in the sequel, § 2.4, where the result of the transformation is not a Markov process.

2.2.3. End of proof

The probability measure P^z_{0, x_0} is absolutely continuous with respect to P_{0, x_0} with a positive density $\pi(0, T)$; so, the converse absolute continuity takes place, and

$$\mathsf{P}_{0, x_0}\{\rho_{0, T}(\xi, \phi) < \delta\} = \mathsf{M}^z_{0, x_0}\{\rho_{0, T}(\xi, \phi) < \delta; \pi(0, T)^{-1}\}. \quad (2.2.13)$$

Take $z(t, x)$ equal to the function mentioned in the formulation of the theorem, and estimate

$$\mathsf{P}^z_{0, x_0}\{\rho_{0, T}(\xi, \phi) < \delta\} \ge 1 - \mathsf{P}^z_{0, x_0}\left\{\sup_{0 \le t \le T} |\xi(t) - \phi(t)| \ge \delta'\right\}.$$

Under the above mentioned choice of $z(t, x)$ the compensator $\tilde{\xi}^z(t)$ turns out to be equal to $\phi(t)$, and Kolmogorov's inequality yields

$$\mathsf{P}^z_{0, x_0}\left\{\sup_{0 \le t \le T} |\xi(t) - \phi(t)| \ge \delta'\right\} \le$$

$$\le (\delta')^{-2} \mathsf{M}^z_{0, x_0} \sum_i \langle \xi^i, \xi^i \rangle^z (T) =$$

$$= (\delta')^{-2} \mathsf{M}^z_{0, x_0} \sum_{t = k\tau = 0}^{T - \tau} \sum_i \frac{\partial^2 G}{\partial z_i^2}(t, \xi(t); z(t, \xi(t))) \cdot \tau \le$$

$$\le (\delta')^{-2} \sum_{t = k\tau = 0}^{T - \tau} D(t) \cdot \tau = \frac{1}{4}.$$

The same is true in the continuous case (if we replace the sum over t by the integral). So

$$\mathsf{P}^z_{0, x_0}\{\rho_{0, T}(\xi, \phi) < \delta'\} \ge \frac{3}{4}.$$

In order to estimate the density $\pi(0, T)^{-1}$ from below, at least on a part of the integration range, we again use Kolmogorov's inequality. Put

$$\zeta(t) = \sum_{s = k\tau = 0}^{t-\tau} z(s, \xi(s))\,(\xi(s+\tau) - \xi(s))$$

or, in the continuous case,

$$\zeta(t) = \int_0^t z(s, \xi(s-))\, d\xi(s).$$

We have

$$\mathsf{P}^z_{0,x_0}\{\zeta(T) - \tilde{\zeta}^z(T) \le -\varepsilon\} \le$$

$$\le \mathsf{P}^z_{0,x_0}\left\{\sup_{0 \le t \le T} |\zeta(t) - \tilde{\zeta}^z(t)| \ge \varepsilon\right\} \le \varepsilon^{-2}\mathsf{M}^z_{0,x_0}\langle\zeta, \zeta\rangle^z(T). \quad (2.2.14)$$

Here the compensator and the quadratic compensator with respect to P^z_{0,x_0} are given by

$$\tilde{\zeta}^z(t) = \sum_{s = k\tau = 0}^{t-\tau} z(s, \xi(s))\,\nabla_z G(s, \xi(s); z(s, \xi(s)))\cdot\tau,$$

$$\langle\zeta, \zeta\rangle^z = \sum_{s = k\tau = 0}^{t-\tau}\sum_{i,j}\frac{\partial^2 G}{\partial z_i \partial z_j}(s, \xi(s); z(s, \xi(s)))\, z_i(s, \xi(s))\, z_j(s, \xi(s))\cdot\tau,$$

respectively, and in the continuous case

$$\tilde{\zeta}^z(t) = \int_0^t z(s, \xi(s))\,\nabla_z G(s, \xi(s); z(s, \xi(s)))\, ds,$$

$$\langle\zeta, \zeta\rangle^z(t) = \int_0^t \sum_{i,j}\frac{\partial^2 G}{\partial z_i \partial z_j}(s, \xi(s); z(s, \xi(s)))\, z_i(s, \xi(s))\, z_j(s, \xi(s))\, ds.$$

If we choose

$$\varepsilon = 2\left(\sum_{t = k\tau = 0}^{T-\tau} ZDZ(t)\cdot\tau\right)^{1/2}$$

(in the continuous case: the same with the integral), then the probability (2.2.14) does not exceed 1/4.

So, with P^z_{0,x_0}-probability not less than 3/4,

$$\pi(0, T)^{-1} > \exp\left\{ \sum_{t=k\tau=0}^{T-\tau} z(t, \xi(t)) \nabla_z G(t, \xi(t); z(t, \xi(t))) \cdot \tau - \right.$$

$$\left. - \sum_{\tau=k\tau=0}^{T-\tau} G(t, \xi(t); z(t, \xi(t))) \cdot \tau - \varepsilon \right\} \qquad (2.2.15)$$

(or, $\pi(0, T)^{-1}$ is greater than the corresponding expression containing the integrals). The probability of the intersection of two events of probability not less than 3/4 is not less than 1/2, that is, the estimate (2.2.15) or its continuous version holds on a part of the integration range in (2.2.13) having $\mathsf{P}^z_{0,\, x_0}$-probability not less than 1/2.

But $z(t, \xi(t)) \nabla_z G(t, \xi(t); z(t, \xi(t))) - G(t, \xi(t); z(t, \xi(t))) = H(t, \xi(t); (\phi(t + \tau) - \phi(t)) / \tau)$ or, in the continuous case, $H(t, x(t); \dot{\phi}(t))$; taking (2.2.4) into account, we obtain (2.2.5), (2.2.5'). ◊

2.3 Derivation of the upper estimate for the probability of going far from the sets $\Phi_{x_0;\,[0, T]}(i)$, $\tilde{\Phi}_{x_0;\,[0, T]}(i)$

2.3.1. Theorem 2.3.1. *Let $(\xi(t), \mathsf{P}_{t,\, x})$ be a Markov process of one of the two classes introduced in* § 2.1; *let the corresponding cumulant satisfy Condition* **A**. *Let δ' and A be arbitrary positive numbers; let $0 = t_0 < t_1 < \dots t_n = T$ be a partition of the interval from 0 to T (in the discrete case, let t_i be multiples of τ),*

$$\Delta t_{\min} = \min_m (t_{m+1} - t_m), \quad \Delta t_{\max} = \max_m (t_{m+1} - t_m);$$

let $z(1), \dots, z(k)$ be r-dimensional vectors belonging to Z; $d(1), \dots, d(k)$ numbers such that $d(j) \geq \overline{G}(z(j))$; $U_0 = \{u : z(j)\, u < d(j),\ 1 \leq j \leq k\}$; and let N be a natural number. Let numbers $\varepsilon_1 \in [0, 1)$, $\varepsilon_2 > 0$ be chosen such that for all $s, t \in [0, T]$, $x, y \in R^r$, $|t - s| < \Delta t_{\max}$, $|x - y| \leq 2\delta'$, and all z,

$$G(t, y; (1 - \varepsilon_1) z) \leq (1 - \varepsilon_1)\, G(s, x; z) + \varepsilon_1 A; \qquad (2.3.1)$$

and such that for all $t \in [0, T)$, $x \in R^r$, there exist $z\{1\}, \dots, z\{N\} \in Z$ (depending on t, x) such that

$$\sup_{u \in U_0} [H(t, x; u) - \max_{1 \leq j \leq N} [z\{j\}\, u - G(t, x; z\{j\})]] \leq \varepsilon_2. \qquad (2.3.2)$$

Then for all $x_0 \in R^r$, $i \geq 0$, $\delta \geq 3\delta'$, $\delta' \geq \Delta t_{\max} \sup\{|u| : u \in U_0\}$,

$$\mathsf{P}_{0,\,x_0}\,\{\rho_{0,\,T}\,(\xi,\,\Phi_{x_0;\,[0,\,T]}\,(i))\geq\delta\}\leq$$

$$\leq 2n\sum_{j=1}^{k}\exp\,\{\Delta t_{\min}\,[\overline{G}\,(z\,(j))-d\,(j)]\}\,+$$

$$+\,N^n\exp\,\{-i+i\cdot\varepsilon_1\,(2-\varepsilon_1)+T\,(A\varepsilon_1\,(2-\varepsilon_1)+\varepsilon_2\,(1-\varepsilon_1))\}. \quad (2.3.3)$$

Note that the maximum in (2.3.2) is nothing but the polyhedron circumscribed from below about the graph of the function H; Condition (2.3.1) is, by virtue of Lemma 1.1.2, equivalent to

$$\sup_{\substack{|t-s|<\Delta t_{\max}\\ |x-y|\leq 2\delta'\\ H\,(s,\,x;\,u)<\infty}}\frac{H\,(t,\,y;\,u)-H\,(s,\,x;\,u)}{A+H\,(s,\,x;\,u)}\leq\frac{\varepsilon_1}{1-\varepsilon_1}. \quad (2.3.4)$$

Proof. Introduce the events

$$A\,(m)=\{\xi\,(t_{m+1})-\xi\,(t_m)\in(t_{m+1}-t_m)\,U_0,$$

$$\sup_{t_m\leq t\leq t_{m+1}}|\,\xi\,(t)-\xi\,(t_m)\,|\leq 2\delta'\},\;\; m=0,\,1,\,\ldots,\,n-1.$$

Let l be the random polygon joining the points $(t_m,\,\xi\,(t_m))$, $m=0,\,1,\,\ldots,\,n$. If all events $A\,(m)$ take place, we have $\rho_{0,\,T}\,(\xi,\,l)<3\delta'$, and so

$$\mathsf{P}_{0,\,x_0}\,\{\rho_{0,\,T}\,(\xi,\,\Phi_{x_0;\,[0,\,T]}\,(i))\geq\delta\}\leq$$

$$\leq\mathsf{P}_{0,\,x_0}\left(\bigcup_{m=0}^{n-1}\overline{A\,(m)}\right)+\mathsf{P}_{0,\,x_0}\left(\bigcap_{m=0}^{n-1}A\,(m)\cap\{l\notin\Phi_{x_0;\,[0,\,T]}\,(i)\}\right). \quad (2.3.5)$$

The first term in (2.3.5) does not exceed the sum of the probabilities of the events $\overline{A\,(m)}$; we estimate these probabilities. To this end we will use the following inequality, which is true for right-continuous Markov processes in metric spaces: if Γ is a subset of the δ'-neighbourhood of a point x, then for $s\leq s'$,

$$\mathsf{P}_{s,\,x}\,\{\xi\,(s')\notin\Gamma\ \text{ or }\sup_{s\leq t\leq s'}\rho\,(\xi\,(t),\,x)>2\delta'\}\leq$$

$$\leq\mathsf{P}_{s,\,x}\,\{\xi\,(s')\notin\Gamma\}+\sup_{\substack{t,\,y\\ s\leq t\leq s'}}\mathsf{P}_{t,\,y}\,\{\rho\,(\xi\,(s'),\,y)>\delta'\}.$$

The proof is practically the same as that of Lemma 6.3 in Dynkin [1]. Using this inequality (with notations changed: $s=t_m$, $s'=t_{m+1}$, $\Gamma=x+(t_{m+1}-t_m)\,U_0$) and the Markov property, we obtain

$$\mathsf{P}_{0,\,x_0}\,(\overline{A\,(m)})=\mathsf{P}_{0,\,x_0}\,\{\xi\,(t_{m+1})-\,\xi\,(t_m)\notin(t_{m+1}-t_m)\,U_0$$

or $\sup_{t_m \le t \le t_{m+1}} \rho(\xi(t), \xi(t_{m+1})) > 2\delta'\} \le$

$$\le \sup_x \mathsf{P}_{t_m, x}\{\xi(t_{m+1}) - x \notin (t_{m+1} - t_m) U_0\} +$$

$$+ \sup_{\substack{t, y \\ t_m \le t \le t_{m+1}}} \mathsf{P}_{t, y}\{|\xi(t_{m+1}) - y| > (t_{m+1} - t_m) \sup\{|u| : u \in U_0\}\} \le$$

$$\le 2 \sup_{\substack{t, x \\ t_m \le t \le t_{m+1}}} \mathsf{P}_{t, x}\{\xi(t_{m+1}) - x \notin (t_{m+1} - t_m) U_0\}. \tag{2.3.6}$$

The probability at the right side of (2.3.6) is estimated using an exponential Chebyshev inequality; it does not exceed

$$\mathsf{P}_{t, x}\left(\bigcup_{j=1}^{k} \{z(j)(\xi(t_{m+1}) - \xi(t)) \ge d(j)(t_{m+1} - t_m)\}\right) \le$$

$$\le \sum_{j=1}^{k} \frac{\mathsf{M}_{t, x} \exp\{z(j)(\xi(t_{m+1}) - \xi(t))\}}{\exp\{d(j)(t_{m+1} - t_m)\}} \le$$

$$\le \sum_{j=1}^{k} \exp\{(t_{m+1} - t_m)[\overline{G}(z(j)) - d(j)]\} \le$$

$$\le \sum_{j=1}^{k} \exp\{\Delta t_{\min} [\overline{G}(z(j)) - d(j)]\}.$$

So, the probability of the union of $\overline{A(m)}$ is estimated from above by the first term in (2.3.3).

To estimate the second summand in (2.3.5) we also use an exponential Chebyshev inequality:

$$\mathsf{P}_{0, x_0}\left(\bigcap_{m=0}^{n-1} A(m) \cap \{l \notin \Phi_{x_0; [0, T]}(i))\}\right) =$$

$$= \mathsf{P}_{0, x_0}\left(\bigcap_{m=0}^{n-1} A(m) \cap \{I_{0, T}(l) > i\}\right) \le$$

$$\le \frac{\mathsf{M}_{0, x_0}\left(\bigcap_{m=0}^{n-1} A(m); \exp\{(1 - \varepsilon_1)^2 I_{0, T}(l)\}\right)}{\exp\{(1 - \varepsilon_1)^2 i\}}. \tag{2.3.7}$$

Recall that $I_{0,T}(l)$ is the sum

$$\sum_{t=k\tau=0}^{T-\tau} H\left(t, l(t); \frac{l(t+\tau)-l(t)}{\tau}\right)\cdot\tau,$$

or the integral

$$\int_0^T H(t, l(t); \dot{l}(t))\,dt.$$

The difference between the argument t here and the t_m nearest to it from the left is less than $\Delta t_{\max}$; if the event $A(m)$ occurs, $|l(t) - \xi(t_m)| \le \delta'$; the third argument in H is equal to $(\xi(t_{m+1}) - \xi(t_m)) / (t_{m+1} - t_m)$. By formula (2.3.4) we obtain

$$I_{0,T}(l) \le \frac{\varepsilon_1}{1-\varepsilon_1} A \sum_{m=0}^{n-1} (t_{m+1} - t_m) +$$

$$+ (1-\varepsilon_1)^{-1} \sum_{m=0}^{n-1} (t_{m+1} - t_m)\, H\left(t_m, \xi(t_m); \frac{\xi(t_{m+1}) - \xi(t_m)}{t_{m+1} - t_m}\right).$$

So we have to estimate

$$\mathsf{M}_{0,x_0}\left(\bigcap_{m=0}^{n-1} A(m); \exp\{(1-\varepsilon_1)\sum_{m=0}^{n-1}(t_{m+1}-t_m)\times\right.$$

$$\left.\times H\left(t_m, \xi(t_m); \frac{\xi(t_{m+1}) - \xi(t_m)}{t_{m+1} - t_m}\right)\}\right).$$

Using the Markov property with respect to the times t_m we find that this expectation does not exceed

$$\prod_{m=0}^{n-1} \sup_x \mathsf{M}_{t_m,x}\left(A(m); \exp\{(t_{m+1} - t_m)(1-\varepsilon_1)\times\right.$$

$$\left.\times H\left(t_m, x; \frac{\xi(t_{m+1}) - x}{t_{m+1} - t_m}\right)\}\right). \qquad (2.3.8)$$

We fix x and proceed to estimate the expectation in (2.3.8) by comparing the function $H(t_m, x; u)$ and its approximating polyhedron (approximation from below). Since the event $A(m)$ occurs, the third argument in H belongs to U_0, and we have:

$$H\left(t_m, x; \frac{\xi(t_{m+1}) - x}{t_{m+1} - t_m}\right) \le$$

$$\le \max_{1 \le j \le N} [z\{j\}(\xi(t_{m+1}) - x)/(t_{m+1} - t_m) - G(t_m, x; z\{j\})] +$$

$$+ \sup_{u \in U_0} \Delta H(t_m, x; u),$$

where

$$\Delta H(t_m, x; u) = H(t_m, x; u) - \max_{1 \le j \le N} [z\{j\} u - G(t_m, x; z\{j\})].$$

Therefore the expectation in (2.3.8) does not exceed

$$\mathsf{M}_{t_m, x}(A(m); \exp\{(t_{m+1} - t_m)(1 - \varepsilon_1) \times$$
$$\times \max_{1 \le j \le N} [z\{j\}(\xi(t_{m+1}) - x)/(t_{m+1} - t_m) - G(t_m, x; z\{j\})] +$$
$$+ (t_{m+1} - t_m)(1 - \varepsilon_1) \sup_{u \in U_0} \Delta H(t_m, x; u)\}) \le$$
$$\le \exp\{(t_{m+1} - t_m)(1 - \varepsilon_1) \cdot \sup_{u \in U_0} \Delta H(t_m, x; u)\} \times$$
$$\times \mathsf{M}_{t_m, x}(A(m); \max_{1 \le j \le N} [\exp\{(1 - \varepsilon_1) z\{j\}(\xi(t_{m+1}) - x) -$$
$$- (t_{m+1} - t_m)(1 - \varepsilon_1) G(t_m, x; z\{j\})\}]).$$

Then we use the fact that $\max_{1 \le j \le N} \le \sum_{j=1}^{N}$ and obtain that the last expectation can be estimated from above by the sum over j from 1 to N of summands of the form

$$\mathsf{M}_{t_m, x}(A(m); \exp\{(1 - \varepsilon_1) z\{j\}(\xi(t_{m+1}) - x) -$$
$$- (t_{m+1} - t_m)(1 - \varepsilon_1) G(t_m, x; z\{j\})\}). \tag{2.3.9}$$

We know that

$$\mathsf{M}_{t_m, x} \exp\{(1 - \varepsilon_1) z\{j\}(\xi(t_{m+1}) - x) -$$
$$- \sum_{t = k\tau = t_m}^{t_{m+1} - \tau} G(t, \xi(t); (1 - \varepsilon_1) z\{j\}) \cdot \tau\} = 1$$

(in the continuous case the integral from t_m to t_{m+1} replaces the sum). Using (2.3.1) we obtain that the expectation (2.3.9) does not exceed

$$\mathsf{M}_{t_m, x}(A(m); \exp\{(1 - \varepsilon_1) z\{j\}(\xi(t_{m+1}) - x) -$$

$$-\sum_{t=k\tau=t_m}^{t_{m+1}-\tau} G(t, \xi(t); (1-\varepsilon_1) z\{j\}) \cdot \tau + A(t_{m+1} - t_m)\}).$$

Taking instead of $A(m)$ the whole space as integration range we find that this expression does not exceed $\exp\{(t_{m+1} - t_m)\, \varepsilon_1 A\}$.

Using inequality (2.3.2) we obtain the following estimate:

$$\mathsf{M}_{t_m, x}\left(A(m); \exp\left\{(t_{m+1} - t_m)(1-\varepsilon_1) H\left(t_m, x; \frac{\xi(t_{m+1}) - x}{t_{m+1} - t_m}\right)\right\}\right) \le$$

$$\le N \exp\{(t_{m+1} - t_m)[A\varepsilon_1 + \varepsilon_2(1-\varepsilon_1)]\};$$

then expression (2.3.8) is estimated by the product of the estimates of the factors, that is $N^n \exp\{T[A\varepsilon_1 + \varepsilon_2(1-\varepsilon_1)]\}$; so

$$\mathsf{M}_{0, x_0}\left(\bigcap_{m=0}^{n-1} A(m); \exp\{(1-\varepsilon_1)^2 I_{0,T}(l)\}\right) \le$$

$$\le N^n \exp\{T(A\varepsilon_1 + \varepsilon_2(1-\varepsilon_1)) + \varepsilon_1(1-\varepsilon_1) AT\} =$$

$$= N^n \exp\{T(A\varepsilon_1(2-\varepsilon_1) + \varepsilon_2(1-\varepsilon_1))\},$$

from which, using (2.3.5) and (2.3.7), we finally deduce (2.3.3). ◊

2.3.2. Generalization and remarks

Now we give a simple generalization of Theorem 2.3.1, which will make it clumsier but more useful for future applications.

Let $\tilde{G}(t, x; z)$ be a function of $t \in [0, T)$, $x, z \in R^r$, measurable with respect to its three arguments, downward convex and lower semicontinuous in the third argument; and let $\tilde{H}(t, x; u) \leftrightarrow \tilde{G}(t, x; z)$ be its Legendre transform. Denote by $\tilde{I}_{0,T}(\phi)$ the functional defined in the discrete case by

$$\tilde{I}_{0,T}(\phi) = \sum_{t=k\tau=0}^{T-\tau} \tilde{H}\left(t, \phi(t); \frac{\phi(t+\tau) - \phi(t)}{\tau}\right)\tau;$$

and in the continuous case, for absolutely continuous $\phi(t)$, by

$$\tilde{I}_{0,T}(\phi) = \int_0^T \tilde{H}(t, \phi(t); \dot{\phi}(t))\, dt,$$

for other ϕ we take $\tilde{I}_{0,T}(\phi) = \infty$. Denote by $\tilde{\Phi}_{x;[0,T]}(i)$ the set of functions ϕ such that $\phi(0) = x$, $\tilde{I}_{0,T}(\phi) \le i$.

Theorem 2.3.2. *Let Z_0 be an arbitrary subset of the set Z; let the conditions of Theorem 2.3.1 be fulfilled with Condition (2.3.1) replaced by the following* *conditions*:

$$\tilde{G}(t, y; (1-\varepsilon_1) z) \le (1-\varepsilon_1)\tilde{G}(s, x; z) + \varepsilon_1 A \tag{2.3.10}$$

for all $s, t \in [0, T)$, $x, y \in R^r$, $|t-s| < \Delta t_{max}$, $|x-y| \le 2\delta'$, *and all* z;

$$G(t, y; (1-\varepsilon_1) z) \le \tilde{G}(t, y; (1-\varepsilon_1) z) + \varepsilon_1 A \tag{2.3.11}$$

for all $t \in [0, T)$, $y \in R^r$ *and all* $z \in Z_0$; *and with Condition* (2.3.2) *changed to*

$$\sup_{u \in U_0} [\tilde{H}(t, x; u) - \max_{1 \le j \le N} [z\{j\} u - \tilde{G}(t, x; z\{j\})]] \le \varepsilon_2, \tag{2.3.12}$$

where $z\{1\}, \ldots, z\{N\} \in Z_0$ *may depend on* t, x.

Then for any $x_0 \in R^r$, $i \ge 0$, $\delta \ge 3\delta'$, $\delta' \ge \Delta t_{max} \sup\{|u| : u \in U_0\}$,

$$\mathsf{P}_{0, x_0}\{\rho_{0,T}(\xi, \Phi_{x_0; [0,T]}(i)) \ge \delta\} \le 2n \sum_{j=1}^{k} \exp\{\Delta t_{min} [\overline{G}(z(j)) - d(j)]\} +$$

$$+ N^n \exp\{-i + i\varepsilon_1 (2-\varepsilon_1) + T(A\varepsilon_1 (3-\varepsilon_1) + \varepsilon_2 (1-\varepsilon_1))\}. \tag{2.3.13}$$

Note that the conditions of this theorem are far less restrictive than those of Theorem 2.3.1; if the latter are fulfilled, the function $H(t, x; u)$ must, for all t, x, be finite for the same values of u. In Theorem 2.3.2 this condition must be fulfilled by the function $\tilde{H}$, and this is no restriction at all on the Markov process under consideration, since we can choose the function $\tilde{H}$ quite independently of the process $\xi(t)$ (of course, if we want to obtain a good estimate (2.3.13), it is necesary to choose this function in some sensible way, but there still remains much freedom in its choice).

The **proof** is carried out exactly in the same way as that of the previous theorem; instead of (2.3.4) we use the inequality

$$\tilde{H}(t, y; u) \le \frac{\varepsilon_1}{1-\varepsilon_1} A + \frac{1}{1-\varepsilon_1} \tilde{H}(s, x; u),$$

deduced from (2.3.10), and instead of (2.3.1), the inequality

$$(1-\varepsilon_1)\tilde{G}(t_m, x; z\{j\}) \ge G(t, \xi(t); (1-\varepsilon_1) z\{j\}) - 2A\varepsilon_1,$$

deduced from (2.3.10), (2.3.11). ◊

Remark to Theorems 2.3.1 and 2.3.2. These theorems remain true if in formulas (2.3.2) and (2.3.12) we take the supremum over $U_0 \cap \overline{U}$ rather than over U_0. Lemma 2.1.2 is used, the rest of the proof is the same.

2.4. The truncated action functional and the estimates associated with it

2.4.1. The times τ_B, τ_V

Let $(\xi(t), \mathsf{P}_{t,x})$, $0 \le t \le T$, be a Markov process on a $C^{(2)}$-manifold X, either with discrete time time taking as its values multiples of τ, or locally infinitely divisible, with compensating operator given by formula (1.3.1). We will assume that the map ψ is continued as a twice differentiable function beyond its domain.

Let B be an open subset of X; V an open subset of $X \times X$ containing the "diagonal" $\{(x, x): x \in X\}$. We introduce the following Markov times: the first exit time $\tau_B = \min \{t \geq 0: \xi(t) \notin B\}$ from B (if $\xi(t)$ does not leave B at all we assume $\tau_B = \infty$) and the time τ_V of the first jump not belonging to V. Namely, in the discrete case

$$\tau_V = \min \{t = k\tau > 0: (\xi(t - \tau), \xi(t)) \notin V\};$$

in the continuous case

$$\tau_V = \min \{t > 0: (\xi(t-), \xi(t)) \notin V\}$$

($\tau_V = \infty$ if there is no such t).

In this section we will derive estimates analogous to those of § 2.2, § 2.3, not on the time interval $[0, T]$ but on $[0, T] \cap [0, \tau_B] \cap [0, \tau_V)$.

2.4.2. The truncated cumulant, the truncated action functional, the semi-metric $\rho_{0, T \wedge \tau_B \wedge \tau_V -}$

Let us fix some map ψ whose domain W contains all points x, y, $x \in B$, $(x, y) \in V$.

We define the *truncated cumulant* $G_V(t, x; z)$, $0 \leq t < T$, $x \in X$, $z \in R^r$. In the discrete case

$$G_V(t, x; z) = \tau^{-1} \ln [\mathsf{M}_{t, x} \{(x, \xi(t + \tau)) \notin V; 1\} + \\ + \mathsf{M}_{t, x} \{(x, \xi(t + \tau)) \in V; \exp \{z(\psi(\xi(t + \tau)) - \psi(x))\}\}];$$

and in the continuous case

$$G_V(t, x; z) = zb_v(t, x) + \frac{1}{2} \sum_{i, j} z_i z_j a^{ij}(t, x) +$$

$$+ \int_{\{y: (x, y) \in V\}} [\exp \{z(\psi(y) - \psi(x))\} - 1 - z(\psi(y) - \psi(x))] \lambda_{t, x}(dy),$$

where

$$b_V(t, x) = b(t, x) - \int_{\{y: (x, y) \notin V\}} [\psi(y) - \psi(x)] \lambda_{t, x}(dy).$$

If $X = R^r$ and if V is all of $R^r \times R^r$, the truncated cumulant turns into the cumulant.

Note that for $x \in B$ the values of $b_V(t, x)$, $G_V(t, x; z)$ do not depend on the way in which the map ψ is continued beyond W.

By $\overline{G}_V(z)$ we denote a non-negative, downward convex function such that $\overline{G}_V(0) = 0$ and $G_V(t, x; z) \leq \overline{G}_V(z)$ for all $t \in [0, T)$, x and z. We assume that $\overline{G}_V(z) < \infty$ for all z.

Now define the function $H_V(t, x; u) \leftrightarrow G_V(t, x; z)$ with arguments $t \in [0, T]$, $x \in X$, $u \in R^r$.

We define the *truncated action functional* $I^V_{0,T}(\phi)$ for functions $\phi(t)$, $0 \leq t \leq T$, with values in the domain of the map. In the discrete case put

$$I^V_{0,T}(\phi) = \sum_{t = k\tau = 0}^{T-\tau} H_V\left(t, \phi(t); \frac{\psi(\phi(t+\tau)) - \psi(\phi(t))}{\tau}\right) \cdot \tau;$$

and in the continuous case, for absolutely continuous functions ϕ,

$$I^V_{0,T}(\phi) = \int_0^T H_V\left(t, \phi(t); \frac{d}{dt} \psi(\phi(t))\right) dt,$$

while for other functions $I^V_{0,T}(\phi) = \infty$.

Note that if the values of a function ϕ lie in the domains of two different maps, the value of the functional $I^V_{0,T}(\phi)$ evaluated with respect to one map is, generally speaking, different from its value with respect to the other map.

We introduce the following semi-metric (depending on the random times τ_B, τ_V):

$$\rho_{0, T \wedge \tau_B \wedge \tau_V -}(\phi, \psi) = \sup_{[0, T] \cap [0, \tau_B] \cap [0, \tau_V)} \rho(\phi(t), \psi(t)).$$

2.4.3. Theorem 2.4.1. *Let $\phi(t)$, $0 \leq t \leq T$, be a function with values in the set B, and let δ be a positive number. Let, in the discrete case, for all x of B and all t, $0 \leq t \leq T$, that are multiples of τ, the function $H_V(t, x; u)$ be differentiable with respect to u at the point $\tau^{-1}(\psi(\phi(t+\tau)) - \psi(\phi(t)))$ and let all values of the function*

$$z(t, x) = \nabla_u H_V\left(t, x; \frac{\psi(\phi(t+\tau)) - \psi(\phi(t))}{\tau}\right)$$

belong to some bounded set; in the continuous case let the same be true for $x \in B$ *and almost all* $t \in [0, T)$ *for the values of* $z(t, x) = \nabla_u H_V\left(t, x; \frac{d}{dt}(\psi(\phi(t)))\right)$.

Take, in the discrete case, for $0 \le t < T$,

$$D(t) = \sup_{x \in B} \sum_i \frac{\partial^2 G_V}{\partial z_i^2}(t, x; z(t, x)), \tag{2.4.1}$$

$$ZDZ(t) = \sup_{x \in B} \sum_{i,j} \frac{\partial^2 G_V}{\partial z_i \partial z_j}(t, x; z(t, x))\, z_i(t, x)\, z_j(t, x), \tag{2.4.2}$$

$$\delta' = 2\left(\sum_{t = k\tau = 0}^{T-\tau} D(t) \cdot \tau\right)^{1/2}, \tag{2.4.3}$$

$$\Delta H(t) = \sup_{\substack{x \in B \\ |\psi(x) - \psi(\phi(t))| < \delta'}} \left[H_V\left(t, x; \frac{\psi(\psi(t+\tau)) - \psi(\phi(t))}{\tau}\right) + \right.$$

$$\left. + H_V\left(t, \phi(t); \frac{\psi(\phi(t+\tau)) - \psi(\phi(t))}{\tau}\right)\right]; \tag{2.4.4}$$

and in the continuous case, the same with $(\psi(\phi(t+\tau)) - \psi(\phi(t)) / \tau$ *in* (2.4.4) *replaced by* $d/dt\, \psi(\phi(t))$, *and the sum in* (2.4.3) *replaced by the integral*

$$\delta' = 2\left(\int_0^T D(t)\, dt\right)^{1/2}. \tag{2.4.3'}$$

Let, for any pair of points x, y *of* B, *if* $\psi(x)$ *and* $\psi(y)$ *are at a distance less than* δ', *then* $\rho(x, y) < \delta$.

Then for $x_0 = \phi(0)$,

$\mathsf{P}_{0, x_0}\{\rho_{0, T \wedge \tau_B \wedge \tau_{V^-}}(\xi, \phi) < \delta\} \ge$

$$\ge \frac{1}{2}\exp\left\{-I^V_{0,T}(\phi) - 2\left(\sum_{t = k\tau = 0}^{T-\tau} ZDZ(t) \cdot \tau\right)^{1/2} - \sum_{t = k\tau = 0}^{T-\tau} \Delta H(t) \cdot \tau\right\} \tag{2.4.5}$$

in the discrete case; in the continuous case,

$$\mathsf{P}_{0,x_0}\{\rho_{0,T\wedge\tau_B\wedge\tau_V-}(\xi,\phi)<\delta\}\geq$$

$$\geq\frac{1}{2}\exp\left\{-I^V_{0,T}(\phi)-2\left(\int_0^T ZDZ(t)\,dt\right)^{1/2}-\int_0^T \Delta H(t)\,dt\right\}. \qquad (2.4.5')$$

It follows from (2.4.5) and (2.4.5') that the probability $\mathsf{P}_{0,x_0}\{\rho_{0,T}(\xi,\phi)<\delta\}$ is estimated from below by the same expression minus $\mathsf{P}_{0,x_0}\{\tau_B\leq T\}$ and $\mathsf{P}_{0,x_0}\{\tau_V\leq T\}$.

Proof. We introduce a truncated Cramér's transformation. To this end we define a random function $\xi_V(t)$ by stopping $\xi(t)$ at the last time before the first jump not belonging to V: in the discrete case

$$\xi_V(t)=\begin{cases}\xi(t) & \text{for } 0\leq t<\tau_V,\\ \xi(\tau_V-\tau) & \text{for } \tau_V\leq t\leq T;\end{cases}$$

in the continuous case τ_V- t is replaced by τ_V-.

For a smooth function $f(t,x)$ the compensator of $f(t,\xi_V(t))$ with respect to the probability P_{0,x_0} is equal to

$$\widetilde{f(t,\xi_V(t))}=f(0,x_0)+\int_0^{t\wedge\tau_V}\mathfrak{A}_V f(s,\xi_V(s))\,ds,$$

$$\mathfrak{A}_V f(t,x)=\frac{\partial f}{\partial t}(t,x)+\nabla_x f(t,x)\cdot b_V(t,x)+\frac{1}{2}\sum_{i,j}a^{ij}(t,x)\frac{\partial^2 f}{\partial x^i\partial x^j}(t,x)+$$

$$+\int_{\{y:\,(x,y)\in V\}}[f(t,y)-f(t,x)-\nabla_x f(t,x)(\psi(y)-\psi(x))]\,\lambda_{t,x}(dy)$$

(for continuous time; in the discrete case the integral is replaced by a sum from 0 to $t\wedge\tau_V-\tau$, and the expression for $\mathfrak{A}_V f$ is changed correspondingly).

We associate to an arbitrary function $z(t,x)$, $0\leq t<T$, $x\in B$, with values in R^r, a functional $\pi_{B,V}(0,t)$ of the paths of the stochastic process $\xi(t)$: in the discrete case

$$\pi_{B,V}(0,t)=\exp\left\{\sum_{s=k\tau=0}^{t\wedge\tau_B\wedge\tau_V-\tau} z(s,\xi(s))(\psi(\xi_V(s+\tau))-\psi(\xi_V(s)))-\right.$$

$$- \sum_{t = k\tau = 0}^{t \wedge \tau_B \wedge \tau_V - \tau} G_V(s, \xi(s); z(s, \xi(s))) \cdot \tau \Bigg\};$$

in the continuous case

$$\pi_{B,V}(0, t) = \exp\Bigg\{ \int_0^t \chi_{[0, \tau_B \wedge \tau_V]}(s)\, z(s, \xi(s-))\, d\psi(\xi_V(s)) -$$

$$- \int_0^{t \wedge \tau_B \wedge \tau_V} G_V(s, \xi(s); z(s, \xi(s)))\, ds \Bigg\}.$$

We verify that $\pi_{B,V}(0, t)$ is a martingale with respect to each of the probabilities P_{0, x_0} and the family of σ-algebras $\mathfrak{F}_{0t}$. In the discrete case this can be done by direct calculation; in the continuous case we evaluate the Lévy measure corresponding to the jumps of the process $\xi_V(t \wedge \tau_B)$:

$$L(A) = \int_0^{T \wedge \tau_B \wedge \tau_V} \lambda_{t, \xi(t)} \{y: (t, y) \in A, (\xi(t), y) \in V\}\, dt,$$

the compensator of $\psi(\xi_V(t \wedge \tau_B))$, which turns out to be

$$\psi(x_0) + \int_0^{t \wedge \tau_B \wedge \tau_V} b_V(s, \xi(s))\, ds,$$

and the bicompensators of the continuous parts of the random functions $\psi^i(\xi_V(t \wedge \tau_B))$, $\psi^j(\xi_V(t \wedge \tau_B))$:

$$\int_0^{t \wedge \tau_B \wedge \tau_V} a^{ij}(s, \xi(s))\, ds,$$

and we use formula (1.2.5).

We obtain that $\mathsf{M}_{0, x_0} \pi_{B,V}(0, T) = 1$. By definition the *truncated Cramér's transformation* associated with a given function $z(t, x)$ replaces the probability measure P_{0,x_0} by the measure P^z_{0, x_0} defined by

$$\mathsf{P}^z_{0, x_0}(A) = \mathsf{M}_{0, x_0}(A; \pi_{B,V}(0, T)).$$

For an arbitrary smooth function $f(t, x)$, we evaluate the compensator of $f(t, \xi_V(t))$ with respect to the probability measure P^z_{0, x_0}.

The generalized Cramér's transformation in § 2.2 maps $\xi(t) - \phi(t)$ to a martingale; the truncated transformation, if $z(t, x)$ is chosen in the way indicated in Theorem 2.4.1, martingalizes the random function

$$\psi(\xi_V(t \wedge \tau_B)) - \psi(\phi(t \wedge \tau_B \wedge \tau_V)).$$

From this point the proof follows that of Theorem 2.2.1; we obtain an estimate for

$$\mathrm{P}_{0, x_0}\left\{\sup_{0 \le t \le T} |\psi(\xi_V(t \wedge \tau_B)) - \psi(\phi(t \wedge \tau_B \wedge \tau_V))| < \delta'\right\}.$$

But the points $\xi_V(t \wedge \tau_B)$ and $\phi(t \wedge \tau_B \wedge \tau_V)$ lie both inside the domain W of our map ψ, and the fact that the Euclidean distance between $\psi(\xi_V(t \wedge \tau_B))$ and $\psi(\phi(t \wedge \tau_B \wedge \tau_V))$ is less than δ' implies that the distance in the manifold between $\xi_V(t \wedge \tau_B)$ and $\phi(t \wedge \tau_B \wedge \tau_V)$ is less than δ. Finally, we take into account the fact that for $t \in [0, \tau_B] \cap [0, \tau_V)$ we have

$$\xi_V(t \wedge \tau_B) = \xi(t), \quad \phi(t \wedge \tau_B \wedge \tau_V) = \phi(t). \ \Diamond$$

2.4.2. Theorem 2.4.2 and a remark

Now let $\tilde{G}(t, x; z)$ be a function of $t \in [0, T)$, $x \in W$, $z \in R^r$, measurable with respect to all three arguments, downward convex and lower semicontinuous with respect to the third argument; let $\tilde{H}(t, x; z)$ be its Legendre transform. For a function $\phi(t)$, $t \in [T_1, T_2] \subseteq [0, T]$, with values in the domain of a chosen map, denote by $\tilde{I}_{T_1, T_2}(\phi)$ the functional defined in the discrete case (for T_1 and T_2 multiples of τ) by the formula

$$\tilde{I}_{T_1, T_2}(\phi) = \sum_{t = k\tau = T_1}^{T_2 - \tau} \tilde{H}\left(t, \phi(t); \frac{\psi(\phi(t + \tau)) - \psi(\phi(t))}{\tau}\right) \cdot \tau;$$

and in the continuous case (for absolutely continuous $\phi(t)$) by

$$\tilde{I}_{T_1, T_2}(\phi) = \int_{T_1}^{T_2} \tilde{H}\left(t, \phi(t); \frac{d}{dt}\psi(\phi(t))\right) dt;$$

for other ϕ we put $\tilde{I}_{T_1, T_2}(\phi) = \infty$. For $[T_1, T_2] \subseteq [0, T]$, $x \in B$ and $i \geq 0$ we denote by $\tilde{\Phi}_{x; [T_1, T_2]}(i)$ the set of functions $\phi(t)$ in the interval from T_1 to T_2 such that $\phi(T_1) = x$, $\tilde{I}_{T_1, T_2}(\phi) \leq i$.

Theorem 2.4.2. *Let the δ-neighbourhood $B_{+\delta}$ of a set B lie within the domain of a given map. Let δ' be an arbitrary positive number less than δ/3; let $A > 0$; let $0 = t_0 < t_1 < \ldots < t_n = T$ be a partition of the interval from 0 to T (in the discrete case, let t_i be multiples of τ), let*

$$\Delta t_{\min} = \min_m (t_{m+1} - t_m), \quad \Delta t_{\max} = \max_m (t_{m+1} - t_m);$$

let $z(1), \ldots, z(k) \in R^r$; let $d(1), \ldots, d(k)$ be numbers such that $d(j) \geq \overline{G}_V(z(j))$; let $U_0 = \{u: z(j)\,u < d(j),\ 1 \leq j \leq k\}$; and let N be a natural number. Let, for x, y in the domain of the given map,

$$|\psi(y) - \psi(x)| < \Delta t_{\max} \sup\{|u| : u \in U_0\}$$

imply $\rho(x, y) < \delta'$. Let numbers $\varepsilon_1 \in [0, 1)$, $\varepsilon_2 > 0$ be such that for all $s, t \in [0, T]$, $x, y \in B_{+\delta}$, $|t - s| < \Delta t_{\max}$, $\rho(x, y) \leq 2\delta'$, and all z the following conditions are satisfied:

$$\tilde{G}(t, y; (1 - \varepsilon_1) z) \leq (1 - \varepsilon_1) \tilde{G}(s, x; z) + \varepsilon_1 A;$$

for all $t \in [0, T)$, $y \in B_{+\delta}$, $z \in Z_0$:

$$G_V(t, y; (1 - \varepsilon_1) z) \leq \tilde{G}(t, y; (1 - \varepsilon_1) z) + \varepsilon_1 A;$$

for all $s \in [0, T)$, $x \in B_{+\delta}$ there exist $z\{1\}, \ldots, z\{N\} \in Z_0$ such that

$$\sup_{u \in U_0} [\tilde{H}(s, x; u) - \max_{1 \leq j \leq N} [z\{j\}\, u - \tilde{G}(s, x; z\{j\})]] \leq \varepsilon_2.$$

Then for any $x_0 \in B$, $i \geq 0$,

$$\mathsf{P}_{0, x_0}\{\rho_{0, T \wedge \tau_B \wedge \tau_V} - (\xi, \tilde{\Phi}_{x_0; [0, T \wedge \tau_B \wedge \tau_V]}(i)) \geq \delta\} \leq$$

$$\leq 2n \sum_{j=1}^{k} \exp\{\Delta t_{\min} [\overline{G}_V(z(j)) - d(j)]\} + N^n \exp\{-i + i\varepsilon_1(2 - \varepsilon_1) +$$

$$+ (1 - \varepsilon_1)^2 \Delta t_{\max} \cdot \sup_{t, x} \tilde{H}(t, x; 0) + T(A\varepsilon_1(3 - \varepsilon_1) + \varepsilon_2(1 - \varepsilon_1))\}. \quad (2.4.6)$$

The estimate (2.4.6) implies that for the probability

$$\mathsf{P}_{0, x_0}\{\rho_{0, T}(\xi, \tilde{\Phi}_{x_0; [0, T]}(i)) \geq \delta\}$$

the same holds, but with $\mathsf{P}_{0,\,x_0}\{\tau_B \le T\} + \mathsf{P}_{0,\,x_0}\{\tau_V \le T\}$ added to the right side.

Proof. Let m_0 be the largest of all numbers m such that $t_m < \tau_B \wedge \tau_V$. Define a random function $l(t)$: on the intervals from t_m to t_{m+1}, $m < m_0$, we define it so that

$$l(t_m) = \xi(t_m),\ l(t_{m+1}) = \xi(t_{m+1})$$

and so that its image on the map is a straight line; for $t_{m_0} \le t \le T \wedge \tau_B \wedge \tau_V$ we put $l(t) = \xi(t_{m_0})$. Introduce the events

$$A(m) = \{\psi(\xi_V(t_{m+1} \wedge \tau_B)) - \psi(\xi_V(t_m \wedge \tau_B)) \in (t_{m+1} - t_m)\, U_0,$$
$$\sup_{t_m \le t \le t_{m+1}} \rho(\xi_V(t \wedge \tau_B), \xi_V(t_m \wedge \tau_B)) \le 2\delta'\},\ \ m = 0, 1, \ldots, n-1.$$

If all events $A(m)$ occur, then $\rho_{0,\,T \wedge \tau_B \wedge \tau_V -}(\xi, l) < \delta$; so,

$$\mathsf{P}_{0,\,x_0}\{\rho_{0,\,T \wedge \tau_B \wedge \tau_V -}(\xi, \tilde{\Phi}_{x_0;\,[0,\,T \wedge \tau_B \wedge \tau_V]}(i)) \ge \delta\} \le$$
$$\le \mathsf{P}_{0,\,x_0}\left(\bigcup_{m=0}^{n-1} \overline{A(m)}\right) + \mathsf{P}_{0,\,x_0}\left(\bigcap_{m=0}^{n-1} A(m) \cap \{\tilde{I}_{0,\,T \wedge \tau_B \wedge \tau_V}(l) > i\}\right).$$

Now use the additivity of the functional $\tilde{I}$:

$$\tilde{I}_{0,\,T \wedge \tau_B \wedge \tau_V}(l) \le \sum_{m=0}^{m_0 - 1} \tilde{I}_{t_m,\,t_{m+1}}(l) + \Delta t_{\max} \cdot \sup_{t,\,x} \tilde{H}(t, x; 0).$$

Then the proof copies, just as the proof of Theorem 2.3.2, that of Theorem 2.3.1. ◊

Remark. Define a distance $\rho_{\{0,\,T\}}$ between functions of the interval from 0 to T as follows: $\rho_{\{0,\,T\}}(\phi, \psi) = \rho_{0,\,T}(\phi, \psi)$ if $\phi(0) = \psi(0)$, $\phi(T) = \psi(T)$; and $\rho_{\{0,\,T\}}(\phi, \psi) = \infty$ otherwise. Then, under the conditions of Theorem 2.4.2,

$$\mathsf{P}_{0,\,x_0}\{\rho_{\{0,\,T\}}(\xi, \tilde{\Phi}_{x_0;\,[0,\,T]}(i)) \ge \delta\}$$

is estimated from above by the right side of (2.4.6) to which $\mathsf{P}_{0,\,x_0}\{\tau_B \le T\} + \mathsf{P}_{0,\,x_0}\{\tau_V \le T\}$ is added.

This follows from the method of proof, which consists of approximating the path by a polygon $l(t)$ coinciding with it for $t = t_0 = 0,\ t_1, \ldots, t_n = T$.

CHAPTER 3

THE ACTION FUNCTIONAL FOR FAMILIES OF MARKOV PROCESSES

3.1. The properties of the functional $S_{T_1, T_2}(\phi)$

3.1.1. Requirements A, B, C

Before we deduce from the estimates of the previous chapter rough limit theorems on large deviations for families of Markov processes, let us study the properties of the functional defined, for absolutely continuous functions, by

$$S_{T_1, T_2}(\phi) = \int_{T_1}^{T_2} H_0(t, \phi(t); \dot{\phi}(t))\, dt \tag{3.1.1}$$

(for other functions S_{T_1, T_2} is taken to be $+\infty$), where $H_0(t, x; u)$ is a function of $t \in [0, T]$, $x, u \in R^r$, lower semicontinuous and downward convex with respect to the third argument. (In the next section, functionals of this form will play the rôle of normalized action functionals for families of processes depending on a parameter.) Put

$$\Phi_{x; [T_1, T_2]}(s) = \{\phi : \phi(T_1) = x, S_{T_1, T_2}(\phi) \le s\}.$$

Consider the Legendre transform $G_0(t, x; z) \leftrightarrow H_0(t, x; u)$. We subject the functions $G_0 \leftrightarrow H_0$ to some restrictions:

A. $G_0(t, x; z) \le \overline{G}_0(z)$ for all t, x, z, where $\overline{G}_0$ is a downward convex non-negative function, finite for all z, and such that $G_0(t, x; 0) \equiv \overline{G}_0(0) = 0$ (compare with Condition **A** of § 2.1).

For the function H_0 this condition means that $\underline{H}_0(u) \le H_0(t, x; u)$ for all t, x, u, where $\underline{H}_0(u) \leftrightarrow \overline{G}_0(z)$; the condition of finiteness of $\overline{G}_0$ for all z becomes

$$\lim_{|u| \to \infty} \underline{H}_0(u) / |u| = \infty,$$

and the condition $\overline{G}_0(0) = 0$ gives $\underline{H}_0(u) \ge 0$.

B. $H_0(t, x; u) < \infty$ for the same u for which $\underline{H}_0(u)$ is finite.

C. $$\Delta H_0(h, \delta') = \sup_{\substack{|t - s| \le h \\ |x - y| \le \delta' \\ H_0(t, x; u) < \infty}} \frac{H_0(s, y; u) - H_0(t, x; u)}{1 + H_0(t, x; u)} \to 0$$

for all $h \downarrow 0$, $\delta' \downarrow 0$.

We will not require that the function G_0 (as is done for the cumulant G in § 2.1) can be expressed in terms of integrals of exponents with respect to some measure; the functions G_0 and H_0 need not be smooth. However, Condition C implies continuity of the function $H_0(t, x; u)$ for u in the set $\{u: H_0(t, x; u) < \infty\}$ and continuity of $G_0(t, x; z)$ for all values of its arguments.

3.1.2. Theorem 3.1.1

We say that functions ϕ of some family of functions are *absolutely equi-continuous* if for any $\delta > 0$ there exists an $\varepsilon > 0$ such that

$$\sum_i |\phi(t_i) - \phi(s_i)| < \delta$$

for any function of the family as soon as the sum of lengths of a finite number of non-overlapping intervals $[s_i, t_i]$ is less than ε. It is easily seen that for absolutely equi-continuity of functions of some family it is necessary and sufficient that these functions are absolutely continuous and that their derivatives are uniformly integrable.

Theorem 3.1.1. *Let the functions* G_0, H_0 *satisfy Conditions* **A** - **C**. *Then*

a) *the functions of the set*

$$\bigcup_{[T_1, T_2] \subseteq [0, T]} \bigcup_x \Phi_{x; [T_1, T_2]}(s)$$

are absolutely equi-continuous;

b) *the functional* $S_{T_1, T_2}(\phi)$ *is lower semicontinuous with respect to uniform convergence; i.e.* $\phi_n \to \phi$ *implies*

$$\varliminf_{n \to \infty} S_{T_1, T_2}(\phi_n) \ge S_{T_1, T_2}(\phi);$$

c) *for any compactum K the set*

$$\bigcup_{x \in K} \Phi_{x; [T_1, T_2]}(s) = \{\phi: \phi(T_1) \in K, S_{T_1, T_2}(\phi) \le s\}$$

is compact with respect to uniform convergence.

Proof. Assertion a) follows from the fact that for ϕ in the set mentioned,

$$\int_{T_1}^{T_2} \underline{H}_0(\dot{\phi})(t))\, dt \le s < \infty,$$

and that $\underline{H}_0(u) / |u| \to \infty$ as $|u| \to \infty$.

Let us prove b). It is sufficient to consider the case when $\lim_{n \to \infty} S_{T_1, T_2}(\phi_n) = s_\infty$ exists. If $s_\infty = \infty$, there is nothing to prove; consider the case $s_\infty < \infty$. We may suppose that $S_{T_1, T_2}(\phi_n) < s_\infty + 1$ for all n; then, by a), the functions ϕ_n are absolutely equi-continuous. Passing to the limit in the inequality

$$\sum_i |\phi_n(t_i) - \phi_n(s_i)| < \delta$$

we obtain that

$$\sum_i |\phi(t_i) - \phi(s_i)| \le \delta \text{ if } \sum_i (t_i - s_i) < \varepsilon,$$

i.e. the function ϕ is absolutely continuous.

Now choose positive h and δ' so that $\Delta H_0(h, \delta')$ is less than a pre-set positive number κ; if necessary, decrease h so that the variations of each of the functions ϕ_n, ϕ on the intervals of length less than h are less than δ'. Consider an arbitrary partition of the interval $[T_1, T_2]$ by points $T_1 = t_0 < t_1 < \dots t_{k-1} < t_k = T_2$ into subintervals of lengths less than h. By the choice of h and δ' we have

$$(1 + \kappa) \int_{T_1}^{T_2} H_0(t, \phi_n(t); \dot\phi_n(t)) + \kappa (T_2 - T_1) \ge$$

$$\ge \sum_{i=0}^{k-1} \int_{t_i}^{t_{i+1}} H_0(t_i, \phi_n(t_i); \dot\phi_n(t))\, dt .$$

Using Jensen's inequality for the downward convex function $H_0(t_i, \phi_n(t_i); u)$ we obtain the inequality

$$(1 + \kappa) S_{T_1, T_2}(\phi_n) + \kappa (T_2 - T_1) \ge$$

$$\ge \sum_{i=0}^{k-1} (t_{i+1} - t_i) H_0\left(t_i, \phi_n(t_i); \frac{\phi_n(t_{i+1}) - \phi_n(t_i)}{t_{i+1} - t_i}\right).$$

Passing to the limit as $n \to \infty$ we obtain, using the semicontinuity of H_0:

$$(1 + \kappa) s_\infty + \kappa (T_2 - T_1) \ge$$

$$\ge \sum_{i=0}^{k-1} (t_{i+1} - t_i) H_0\left(t_i, \phi(t_i); \frac{\phi(t_{i+1}) - \phi(t_i)}{t_{i+1} - t_i}\right).$$

Using the choice of h and δ' once again, we have

$$H_0(t, \phi(t); u) \le (1 + \kappa) H_0(t_i, \phi(t_i); u) + \kappa$$

for $|t - t_i| < h$, $|\phi(t) - \phi(t_i)| < \delta'$; therefore

$$\sum_{i=0}^{k-1} \int_{t_i}^{t_{i+1}} H_0\left(t, \phi(t); \frac{\phi(t_{i+1}) - \phi(t_i)}{t_{i+1} - t_i}\right) dt \leq$$

$$\leq \kappa (T_2 - T_1) + (1 + \kappa)\, \kappa (T_2 - T_1) + (1 + \kappa)^2 s_\infty.$$

Consider the function defined by

$$\psi(t) = \frac{\phi(t_{i+1}) - \phi(t_i)}{t_{i+1} - t_i}$$

for $t_i < t < t_{i+1}$, $i = 0, 1, \ldots, k-1$. For any sufficiently small partition the following inequality holds:

$$\int_{T_1}^{T_2} H_0(t, \phi(t);\ \psi(t))\, dt \leq (1 + \kappa)^2 s_\infty + (2\kappa + \kappa^2)(T_2 - T_1).$$

For a sequence of partitions becoming indefinitely fine, the corresponding functions $\psi_n(t)$ converge to $\dot{\phi}(t)$ almost everywhere. The function H_0 is lower semicontinuous, hence almost everywhere

$$H_0(t, \phi(t); \dot{\phi}(t)) \leq \underline{\lim}_{n \to \infty}\ H_0(t, \phi(t); \psi_n(t));$$

recalling that $H_0 \geq 0$ and using Fatou's lemma we obtain

$$S_{T_1, T_2}(\phi) = \int_{T_1}^{T_2} H_0(t, \phi(t); \dot{\phi}(t))\, dt \leq \underline{\lim}_{n \to \infty} \int_{T_1}^{T_2} H_0(t, \phi(t); \psi_n(t))\, dt \leq$$

$$\leq (1 + \kappa)^2 s_\infty + (2\kappa + \kappa^2)(T_2 - T_1).$$

Since $\kappa > 0$ is arbitrarily small, we obtain $S_{T_1, T_2}(\phi) \leq s_\infty$, which was to be proved.

b) implies that the set $\bigcup_{x \in K} \Phi_{x; [T_1, T_2]}(s)$ is closed, absolute equicontinuity implies equicontinuity, and the use of Arzela's theorem gives c). ◊

From b) and c) it follows, in particular, that the functional S_{T_1, T_2} attains its minimum value on any non-empty closed set of functions ϕ such that the set of their initial values $\{\phi(T_1)\}$ is bounded.

Assertions b) and c) are part of the definition of the action functional (see Introduction).

3.1.3. Theorem 3.1.2. *Let Conditions* **A** - **C** *be satisfied. Then for any* $\gamma > 0$ *and any* $s_0 > 0$ *there exists an* $h > 0$ *such that for any partition of the interval from* T_1

to T_2 *by points* $T_1 = t_0 < t_1 < \ldots < t_{n-1} < t_n = T_2$, $\max (t_{j+1} - t_j) \le h$, *we have*

$$\text{a)} \qquad \sum_{j=0}^{n-1} (t_{j+1} - t_j)\, H_0 \left(t_j, \phi(t_j); \frac{\phi(t_{j+1}) - \phi(t_j)}{t_{j+1} - t_j} \right) \le$$

$$\le \int_{T_1}^{T_2} H_0 (t, \phi(t); \dot{\phi}(t))\, dt + \gamma \tag{3.1.2}$$

for any absolutely continuous function $\phi(t)$, $T_1 \le t \le T_2$, *such that the integral at the right side does not exceed* s_0;

$$\text{b)} \qquad \int_{T_1}^{T_2} H_0 (t, l(t); \dot{l}(t))\, dt \le$$

$$\le \sum_{j=0}^{n-1} (t_{j+1} - t_j)\, H_0 \left(t_j, l_j; \frac{l_{j+1} - l_j}{t_{j+1} - t_j} \right) + \gamma, \tag{3.1.3}$$

where $l(t)$, $T_1 \le t \le T_2$, *is the polygon with vertices* (t_j, l_j), *and* l_j, $0 \le j \le n$, *are arbitrary elements of* R^r *such that the sum at the right side of* (3.1.3) *does not exceed* s_0.

Proof. For an arbitrary $\kappa > 0$ take the corresponding positive h and δ' (see the proof of the previous theorem); diminish h, if necessary, so that $|\phi(t) - \phi(s)| < \delta'$ for $|t - s| < h$. By Jensen's inequality,

$$\sum_{j=0}^{n-1} (t_{j+1} - t_j)\, H_0 \left(t_j, \phi(t_j); \frac{\phi(t_{j+1}) - \phi(t_j)}{t_{j+1} - t_j} \right) \le$$

$$\le \sum_{j=0}^{n-1} \int_{t_j}^{t_{j+1}} H_0 (t_j, \phi(t_j); \dot{\phi}(t))\, dt \le$$

$$\le \kappa (T_2 - T_1) + (1 + \kappa) \int_{T_1}^{T_2} H_0 (t, \phi(t); \dot{\phi}(t))\, dt .$$

Since κ is arbitrary, we obtain assertion a) of the theorem. The proof of b) is still simpler (one need not even use Jensen's inequality); the h chosen is suitable for $l(t)$ as well. ◊

3.2. Theorems on the action functional for families of Markov processes in R^r. The case of finite exponential moments

3.2.1. Requirements D - E; notations associated with the parameter θ

We impose some more restrictions on the functions $G_0 \leftrightarrow H_0$.

D. The set $\{u: \underline{H}_0\,(u) < \infty\}$ is open, and $\sup\limits_{t,\,x} H_0\,(t, x; u_0) < \infty$ for some point u_0 of it.

Condition **D** excludes some of the examples given in Subsection 2.4.1.

E. For any compactum $U_K \subset \{u: \underline{H}_0\,(u) < \infty\}$ the gradient $\nabla_u H_0\,(t, x; u)$ is bounded and continuous in $u \in U_K$, uniformly with respect to all t, x.

We will prove rough limit theorems on large deviations for families of Markov processes depending on a parameter θ changing in a set on which a filter $\theta \to$ is given (see Introduction, 0.1). We will denote everything involving the process ξ^0 with superscript θ: the cumulant $G^\theta\,(t, x; z)$, its Legendre transform $H^\theta\,(t, x; u)$, the action functional $I^\theta_{T_1,\,T_2}$, etc.

3.2.2. The case of continuous time

Let $(\xi^\theta\,(t), \mathrm{P}^\theta_{t,\,x})$, $0 \le t \le T$, for every $\theta \in \Theta$ be a locally infinitely divisible process in R^r.

Theorem 3.2.1. *Let for some pair of functions* $G_0 \leftrightarrow H_0$ *Conditions* **A** - **E** *be satisfied; let the functional* $S_{0,\,T}$ *be defined by formula* (3.1.1). *Let* $k\,(\theta)$ *be a real-valued function tending to* $+\infty$ *as* $\theta \to$; *let for the cumulant* $G^\theta\,(t, x; z)$ *the following conditions be satisfied*:

$$k\,(\theta)^{-1}\, G^\theta\,(t, x; k\,(\theta)\, z) \to G_0\,(t, x; z), \tag{3.2.1}$$

$$\nabla_z\,(k\,(\theta)^{-1}\, G^\theta\,(t, x; k\,(\theta)\, z)) \to \nabla_z\, G_0\,(t, x; z) \tag{3.2.2}$$

as $\theta \to$, *uniformly with respect to* t, x *and with respect to every bounded set of values of* z; *and, for every bounded set* K, *let*

$$\left| \frac{\partial^2}{\partial z_i \partial z_j}\,(k\,(\theta)^{-1}\, G^\theta\,(t, x; k\,(\theta)\, z)) \right| \le \mathrm{const} < \infty \tag{3.2.3}$$

for all sufficiently far θ, *for all* t, x *and* $z \in K$.

Then for any $\delta > 0$, $\gamma > 0$ *and* $s_0 > 0$, *for sufficiently far* θ, *for all* $x_0 \in R^r$, *and* $\phi \in \Phi_{x_0;\,[0,T]}(s_0)$,

$$\mathsf{P}^{\theta}_{0,x_0}\{\rho_{0,T}(\xi^{\theta},\phi) < \delta\} \geq \exp\{-k(\theta)[S_{0,T}(\phi) + \gamma]\}. \qquad (3.2.4)$$

Proof. Suppose that the point u_0 mentioned in Condition **D** is 0. This is no restriction of generality: we can consider the process $\xi^{\theta}(t) - u_0 t$.

Formulas (3.2.1), (3.2.2) and Condition **E** imply that

$$H^{\theta}(t,x;u) = k(\theta)[H_0(t,x;u) + o(1)],$$

$$\nabla_u H^{\theta}(t,x;u) = O(k(\theta))$$

as $\theta \to$, uniformly with respect to t, x and u in each compactum $U_K \subset \{u: \underline{H}_0(u) < \infty\}$.

We first give the proof for a set of functions ϕ such that all values of $\dot{\phi}(t)$, $0 \leq t \leq T$, belong to one and the same compact set $U_K \subset \{u: \underline{H}_0(u) < \infty\}$. For such functions,

$$I^{\theta}_{0,T}(\phi) = \int_0^T H^{\theta}(t,\phi(t);\ \dot{\phi}(t))\,dt = k(\theta)[S_{0,T}(\phi) + o(1)]$$

as $\theta \to$, uniformly over the whole set.

Use Theorem 2.2.1. Put $z^{\theta}(t,x) = \nabla_u H^{\theta}(t,x;\dot{\phi}(t))$ and estimate the corresponding quantities (2.2.1) - (2.2.4):

$$D^{\theta}(t) = \sup_x \sum_i \frac{\partial^2 G^{\theta}}{\partial z_i^2}(t,x;z^{\theta}(t,x)) = O(k(\theta)^{-1}),$$

$$(ZDZ)^{\theta}(t) = \sup_x \sum_{i,j} \frac{\partial^2 G^{\theta}}{\partial z_i \partial z_j}(t,x;z^{\theta}(t,x))\, z_i^{\theta}(t,x)\, z_j^{\theta}(t,x) = O(k(\theta)),$$

$$(\delta')^{\theta} = 2\left(\int_0^T D^{\theta}(t)\,dt\right)^{1/2} = O(k(\theta)^{-1/2}),$$

$$\Delta H^{\theta}(t) = \sup_{|x-\phi(t)|<(\delta')^{\theta}} [H^{\theta}(t,x;\dot{\phi}(t)) - H^{\theta}(t,\phi(t);\dot{\phi}(t))] \leq$$

$$\leq k(\theta) \sup_{|x-\phi(t)|<(\delta')^{\theta}} [H_0(t,x;\dot{\phi}(t)) - H_0(t,\phi(t);\dot{\phi}(t))] + o(k(\theta)) \leq$$

$$\leq k(\theta)[1 + H_0(t,\phi(t);\dot{\phi}(t))]\,\Delta H_0(0,(\delta')^{\theta}) + o(k(\theta)).$$

For sufficiently far θ we have $(\delta')^{\theta} \le \delta$; and then

$$P^{\theta}_{0, x_0} \{\rho_{0,T}(\xi^{\theta}, \phi) < \delta\} \ge$$

$$\ge \exp\left\{ - I^{\theta}_{0,T}(\phi) - \ln 2 - 2\left(\int_0^T (ZDZ)^{\theta}(t)\, dt \right)^{1/2} - \int_0^T \Delta H^{\theta}(\dot{t})\, dt \right\} =$$

$$= \exp\{- k(\theta) S_{0,T}(\phi) - o(k(\theta))[1 + S_{0,T}(\phi)]\}.$$

For sufficiently far θ the term $\gamma k(\theta)$ compensates the term $o(k(\theta))$, and we obtain (3.2.4).

Now, let ϕ be an arbitrary function with $S_{0,T}(\phi) \le s_0$. Using Theorems 3.1.1 and 3.1.2 we choose $0 = t_0 < t_1 < \dots t_n = T$ (one and the same for all ϕ with $S_{0,T}(\phi) \le s_0$) and for each of the functions ϕ we form the polygon l with vertices $(t_i, \phi(t_i))$, so that

$$\rho_{0,T}(l, \phi) < \delta/3, \quad S_{0,T}(l) \le S_{0,T}(\phi) + \gamma/3.$$

Then we put

$$l_{\kappa}(t) = \kappa x_0 + (1 - \kappa)\, l(t)$$

where κ is a small positive number (and $x_0 = \phi(0) = l(0) = l_{\kappa}(0)$). The values of $\dot{l}_{\kappa}(t)$ belong to some compact subset of $\{u: \underline{H}_0(u) < \infty\}$. Since the distance $\sup_{0 \le t \le T} |\phi(t) - \phi(0)|$ does not exceed some constant for all functions ϕ under consideration, we can make the polygons l and l_{κ} arbitrarily close to each other by choosing κ sufficiently small; in particular, we can make $\rho_{0,T}(\phi, l_{\kappa}) < 2\delta/3$. Now we estimate $S_{0,T}(l_{\kappa})$ using the convexity of the function H_0:

$$\int_0^T H_0(t, l_{\kappa}(t); \dot{l}_{\kappa}(t))\, dt = \int_0^T H_0(t, l_{\kappa}(t); (1 - \kappa)\, \dot{l}(t))\, dt \le$$

$$\le \int_0^T \kappa H_0(t, l_{\kappa}(t); 0)\, dt + \int_0^T (1 - \kappa) H_0(t, l_{\kappa}(t); \dot{l}(t))\, dt.$$

The first integral does not exceed $\kappa T \sup_{t,x} H_0(t, x; 0)$, the second can be made to differ arbitrarily little from $S_{0,T}(l)$; so, finally, for sufficiently small κ and for all ϕ with $S_{0,T}(\phi) \le s_0$ we have: $S_{0,T}(l_{\kappa}) \le S_{0,T}(\phi) + 2\gamma/3$. For the functions l_{κ} inequality (3.2.4) with δ replaced by $\delta/3$ and γ by $\gamma/3$ holds; so, for sufficiently far θ,

$$\mathsf{P}^{\theta}_{0,x_0}\{\rho_{0,T}(\xi^{\theta},\phi)<\delta\}\geq \mathsf{P}^{\theta}_{0,x_0}\{\rho_{0,T}(\xi^{\theta},l_\kappa)<\delta/3\}\geq$$

$$\geq \exp\{-k(\theta)[S_{0,T}(\phi)+\gamma]\}.\ \Diamond$$

3.2.3. Theorem 3.2.2. *Under the conditions of Theorem* 3.2.1, *for any* $\delta>0$, $\gamma>0$, $s_0>0$, *for sufficiently far* θ, *for all* $x\in R^r$ *and* $s\leq s_0$,

$$\mathsf{P}^{\theta}_{0,x_0}\{\rho_{0,T}(\xi^{\theta},\Phi_{x_0;[0,T]}(s))\geq\delta\}\leq \exp\{-k(\theta)(s-\gamma)\}. \qquad (3.2.5)$$

Proof. Again, without loss of generality, we assume $u_0=0$; we use Theorem 2.3.2. We choose the elements of the construction of this theorem in the following order (dependence on the parameter θ is indicated by superscript): the functions $\tilde{G}^{\theta}(t,x;z)\leftrightarrow\tilde{H}^{\theta}(t,x;u)$, the functional $\tilde{I}^{\theta}_{0,T}$ and the set $\tilde{\Phi}^{\theta}_{x_0;[0,T]}(i)$; i^{θ} and A^{θ}; ε_1, δ' and h; $z^{\theta}(j)$, $d^{\theta}(j)$ and the partition $\{t_m\}$; ε_2^{θ}, N and $z^{\theta}\{1\}$, ..., $z^{\theta}\{N\}$; the set Z_0^{θ}. We have to verify that the conditions of the theorem are satisfied for all sufficiently far θ.

Take $\tilde{G}^{\theta}(t,x;z)=k(\theta)\,G_0(t,x;k(\theta)^{-1}z)$; then

$$\tilde{H}^{\theta}(t,x;u)=k(\theta)\,H_0(t,x;u),$$

$$\tilde{I}^{\theta}_{0,T}(\phi)=k(\theta)\,S_{0,T}(\phi),$$

$$\tilde{\Phi}^{\theta}_{x_0;[0,T]}(k(\theta)\,s)=\tilde{\Phi}^{\theta}_{x_0;[0,T]}(s).$$

For i^{θ} we take $k(\theta)\,s$; put $A^{\theta}=k(\theta)$. Then we put

$$\Delta_1=\gamma/4\,(T+s_0),\ \ \varepsilon_1=\Delta_1/(1+\Delta_1).$$

Choose positive numbers $\delta'\leq\delta/3$ and h in such a way that for $|t-s|\leq h$, $|x-y|\leq 2\delta'$ and all u,

$$H_0(s,y;u)-H_0(t,x;u)\leq[1+H_0(t,x;u)]\,\Delta_1. \qquad (3.2.6)$$

If we choose $\{t_m\}$ so that $\Delta t_{\max}\leq h$, Condition (2.3.10) of Theorem 2.3.2 holds.

The partition $\{t_m\}$ will consist of subintervals of identical length $\Delta t\leq h$. Take the vectors $z^{\theta}(j)$ and numbers $d^{\theta}(j)$ of the form $z^{\theta}(j)=k(\theta)\,z_0(j)$, $d^{\theta}(j)=k(\theta)\,d_0(j)$; then the sum over j in the estimate (2.3.3) consists of terms

$$\exp\{\Delta t\,[\overline{G}^{\theta}(k(\theta)\,z_0(j)) - k(\theta)\,d_0(j)]\} =$$

$$= \exp\{k(\theta)\,\Delta t\,[\overline{G}_0(z_0(j)) + o(1) - d_0(j)]\}. \qquad (3.2.7)$$

We use a part of the $z_0(j)$, $d_0(j)$ in order to ensure the inequality $\Delta t \times \sup\{|u|: u \in U_0\} \le \delta'$. Put $z_0(2i-1) = Ze_i$, $z_0(2i) = -Ze_i$, $i = 1, \ldots, r$, where $Z > 0$ and e_i is the unit vector along the i-th coordinate axis; $d_0(1), \ldots, d_0(2r)$ are taken equal to $D > 0$. Then U_0 is contained in a cube with centre at 0 and side $2D/Z$; in order that this cube is contained in the ball of radius $\delta'/\Delta t$ with centre at 0, we relate Z and D by $Z = \Delta t D\sqrt{r}/\delta'$. In order to be able to disregard the summands (3.2.7), $j = 1, \ldots, 2r$, for far values of θ, we restrict still further the choice of Z, D and Δt: we postulate that $\Delta t\,\overline{G}_0(z_0(j))$, $1 \le j \le 2r$, should not exceed s_0, and $\Delta t d_0(j) = \Delta t D = 3s_0$. This gives

$$Z = 3s_0\sqrt{r}/\delta'.$$

So the vectors $z_0(1), \ldots, z_0(2r)$ are defined; for Δt we can take any positive number not exceeding $h \wedge \min_{1 \le j \le 2r} [\overline{G}_0(z(j))]^{-1}$ such that $T/\Delta t$ is an integer. Finally, put $d_0(1) = \ldots = d_0(2r) = D = 3s_0/\Delta t$.

We supplement the $z_0(1), \ldots, z_0(2r)$, $d_0(1), \ldots, d_0(2r)$, already chosen, if necessary, by some more $z_0(j)$, $d_0(j)$ in order to cut from the set U_0 those points u at which the function H_0 is infinite - otherwise Condition (2.3.12) can not hold. To this end we consider the set $\{u: \underline{H}_0(u) \le 2s_0/\Delta t\}$; this set is compact by virtue of the semicontinuity of $\underline{H}_0$ and the fact that $\lim_{|u| \to \infty} \underline{H}_0(u) = \infty$. There exists a positive ρ such that the ρ-neighbourhood of the set $\{u: \underline{H}_0(u) \le 2s_0/\Delta t\}$ is contained in the set $\{u: \underline{H}_0(u) < \infty\}$.

Circumscribe about the $\rho/2$-neighbourhood of the set $\{u: \underline{H}_0(u) \le 2s_0/\Delta t\}$ a polyhedron entirely contained in the ρ-neighbourhood of this set. On both sides of each face $\{u: zu = d\}$ of this polyhedron there are points with $\underline{H}_0(u) < \infty$, and by Lemma 1.1.1 there exists a constant c such that

$$cd - \overline{G}_0(cz) = \inf\{\underline{H}_0(u): zu = d\} > 2s_0/\Delta t.$$

It is clear that $cd > 0$. The above-introduced polyhedron lies entirely on one side of the hyperplane $\{u: czu = cd\}$, namely on the side where $czu \le cd$ (because the point $u = 0$ lies at this side). Take the vectors cz used in the equations of these hyperplanes as

$z_0\,(2r+1), \ldots, z_0\,(k)$, and the numbers cd, as $d_0\,(2r+1), \ldots, d_0\,(k)$.

After supplementing $z_0\,(1), \ldots, z_0\,(2r)$, $d_0\,(1), \ldots, d_0\,(2r)$ by $z_0\,(j)$, $d_0\,(j)$, $2r+1 \le j \le k$, the inequality $\Delta t \sup \{\, |u| : u \in U_0\} \le \delta'$ remains true, and for the $(2r+1)$-th, ..., k-th terms of the sum in (2.3.13) we have:

$$\exp\{\Delta t\,[\overline{G}^{\theta}\,(z^{\theta}\,(j)) - d^{\theta}\,(j)]\} =$$

$$= \exp\{k\,(\theta)\,\Delta t\,[\overline{G}_0\,(z_0\,(j)) + o\,(1) - d_0\,(j)]\} < \exp\{-k\,(\theta)\,s_0\}$$

for sufficiently far θ.

3.2.4. Approximating H_0 by a polyhedron. Completion of the proof

For ε_2^{θ} we take $2k\,(\theta)\,\Delta_1 = k\,(\theta)\,\gamma/2\,(T + s_0)$. We have to choose a number N and for any t, x to assign $z_0\,\{1\}, \ldots, z_0\,\{N\}$ such that the difference between the function $H_0\,(t, x; u)$ and the polyhedron circumscribed about it from below,

$$\max_{1 \le j \le N}\; [z_0\,\{j\}\,u - G_0\,(t, x; z_0\,\{j\})],$$

can be estimated from above on the set U_0 by the quantity $2\Delta_1$. Then Condition (2.3.12) will be satisfied for $z^{\theta}\,\{j\} = k\,(\theta)\,z_0\,\{j\}$.

By Condition **E**, for u belonging to the closure of U_0 we have

$$|\nabla_u H_0\,(t, x; u)| \le C < \infty.$$

Condition **A** implies that

$$|\nabla_z G_0\,(t, x; z)| \le C_1 < \infty,\quad H_0\,(t, x; \nabla_z G_0\,(t, x; z)) \le C_2 \text{ for } |z| \le C.$$

Introduce the compact set

$$U_{C_2} = \{u\colon \underline{H}_0\,(u) \le C_2\}.$$

For $u \in U_{C_2}$ and all t, x the functions $H_0\,(t, x; u)$ are equicontinuous in u (by Condition **E**) and uniformly bounded (by Conditions **D**, **E**); therefore we can choose a finite Δ_1-net $\{H_0\,(s_i, x_i; u),\ i = 1, \ldots, I\}$ in the set of functions $H_0\,(t, x; u)$, $u \in U_{C_2}$. Approximate each of the continuous, downward convex functions $H_0\,(s_i, x_i; u)$ on the set U_0 by a circumscribed polyhedron made up of the planes

$$H_{ij}\,(u) = z_0^i\,\{j\}\,u - G_0\,(s_i, x_i; z_0^i\,\{j\}),\ j = 1, \ldots, N_i,$$

with points of contact $u\,(i, j) = \nabla_z G_0\,(s_i, x_i; z_0^i\,\{j\} \in U_0$ (and $|z_0^i\,\{j\}| \le C_1$); let the accuracy of the approximation be Δ_1.

For each pair t, x we choose s_i, x_i such that

$$\max_{u \in U_{C_2}} |H_0(t, x; u) - H_0(s_i, x_i; u)| < \Delta_1.$$

Then

$$H_{ij}(u) \le H_0(s_i, x_i; u) < H_0(t, x; u) + \Delta_1$$

for all $u \in U_{C_2}$, and for $u = u(i, j)$ we have

$$H_{ij}(u(i, j)) = H_0(s_i, x_i; u(i, j)) > H_0(t, x; u(i, j)) - \Delta_1.$$

Therefore we can shift the plane $H_{ij}(u)$ by a distance at most Δ_1 so that it touches $H_0(t, x; u)$ at a point $u = \nabla_z G_0(t, x; z_0^i\{j\}) \in U_{C_2}$, and so that

$$\sup_{u \in U_0} [H_0(t, x; u) - \max_{1 \le j \le N} [z_0^i\{j\} u - G_0(t, x; z_0^i\{j\})]] \le 2\Delta_1.$$

Now if we take

$$Z_0^\theta = \{z : |z| \le k(\theta) C_1\}$$

and, for arbitrary t, x, put $z^\theta\{j\} = k(\theta) z_0^i\{j\}$, then Condition (2.3.12) is satisfied with $\varepsilon_2^\theta = 2k(\theta)\Delta_1$. For N we take $\max_i N_i$.

The fact that (2.3.11) is satisfied for far values of θ follows from the uniform convergence (3.2.1). As the result we have:

$$\mathsf{P}_{0, x_0}^\theta \{\rho_{0, T}(\xi^\theta, \Phi_{x_0; [0, T]}(s)) \ge \delta\} \le 2nk \exp\{-k(\theta) s_0\} +$$

$$+ N^n \exp\left\{-k(\theta) s + k(\theta) \frac{\gamma}{4(T + s_0)} \cdot (4T + 2s_0)\right\}.$$

For sufficiently far θ the difference between γ and the factor in front of $k(\theta)$ in the second term compensates the first summand and the factor N^n, and we obtain the estimate (3.2.5) for sufficiently far θ, all x_0 and all $s \le s_0$.

Theorem 3.2.2 is proved. ◊

3.2.5. Theorem 3.2.3.

We combine Theorems 3.1.1, 3.2.1 and 3.2.2 to obtain the following theorem:

Theorem 3.2.3. *Under the conditions of Theorem* 3.2.1, $k(\theta) S_{0, T}(\phi)$ *is the action functional for the family of Markov processes* $(\xi^\theta(t), \mathsf{P}_{t, x}^\theta)$ *as* $\theta \to$, *uniformly with respect to the initial point.*

3.2.6. The case of discrete time

Now, let a positive function $\tau(\theta)$ be given on the parameter set Θ. Let $(\xi^\theta(t), \mathsf{P}^\theta_{t,x})$, $0 \le t \le T$, for each $\theta \in \Theta$ be a $\tau(\theta)$-process in R^r (see § 1.2). Denote by $G^\theta(t, x; z)$ the cumulant of the discrete-time Markov process $(\xi^\theta(t), \mathsf{P}^\theta_{t,x})$, $t = k\tau(\theta) \in [0, T]$.

Theorem 3.2.3'. *Let Conditions* **A** - **E** *be fullfilled for some functions* $G_0 \leftrightarrow H_0$, *let the functional* $S_{0,T}$ *be given by formula* (3.1.1); *suppose* $k(\theta) \to +\infty$ *and* $\tau(\theta) \to 0$ *as* $\theta \to$; *let for the cumulant* $G^\theta(t, x; z)$ *Conditions* (3.2.1) - (3.2.3) *be satisfied.*

Then $k(\theta) S_{0,T}(\phi)$ *is the action functional for the family of Markov processes* $(\xi^\theta(t), \mathsf{P}^\theta_{t,x})$ *as* $\theta \to$, *uniformly with respect to the initial point.*

Proof. The part dealing with the normalized action functional has already been proved; we have to prove the inequalities (3.2.4), (3.2.5). Let us prove (3.2.4) at first for piecewise linear functions $\phi(t)$ whose derivative is $\dot\phi(t) \in U_K \subset \{u: \underline{H}_0(u) < \infty\}$ and which remains constant on the intervals $(k\tau(\theta), (k+1)\tau(\theta))$. The action functional $I^\theta_{0,T}(\phi)$ is given by

$$I^\theta_{0,T}(\phi) =$$

$$= \sum_{k=0}^{[T/\tau(\theta)]-1} H^\theta\left(k\tau(\theta), \phi(k\tau(\theta)); \frac{\phi((k+1)\tau(\theta)) - \phi(k\tau(\theta))}{\tau(\theta)}\right) \cdot \tau(\theta) =$$

$$= k(\theta)\left[\sum H_0\left(k\tau(\theta), \phi(k\tau(\theta)); \frac{\phi((k+1)\tau(\theta)) - \phi(k\tau(\theta))}{\tau(\theta)}\right) \cdot \tau(\theta) + o(1)\right].$$

If $S_{0,T}(\phi) \le s_0$, by Theorem 3.1.2 a), for sufficiently far θ (and therefore, small $\tau(\theta)$) the value of $I^\theta_{0,T}(\phi)$ does not exceed $k(\theta)[S_{0,T}(\phi) + \gamma/2]$. Then we use the fact that for sufficiently far θ the variation of the function ϕ over all intervals of length $\tau(\theta)$ is at most $\delta/2$; so

$$\rho_{0,T}(\xi^\theta, \phi) \le \max_{0 \le k \le [T/\tau(\theta)]} |\xi^\theta(k\tau(\theta)) - \phi(k\tau(\theta))| + \delta/2.$$

Now we use the estimate (2.2.5) with $\delta/2$ instead of δ. We have

$$z^\theta(t, x) = \nabla_u H_0\left(t, x; \frac{\phi(t + \tau(\theta)) - \phi(t)}{\tau(\theta)}\right) = O(k(\theta)),$$

$$D^{\theta}(t) = O(k(\theta)^{-1}), \quad (ZDZ)^{\theta}(t) = O(k(\theta)),$$

$$(\delta')^{\theta} = 2\left(\sum_{k=0}^{[T/\tau(\theta)]-1} D^{\theta}(k\tau(\theta)) \cdot \tau(\theta)\right)^{1/2} = O(k(\theta)^{-1/2}),$$

$$\Delta H^{\theta}(t) \le \Delta H_0(0, (\delta')^{\theta}) \times$$

$$\times k(\theta)\left[1 + H_0\left(t, \phi(t); \frac{\phi(t+\tau(\theta)) - \phi(t)}{\tau(\theta)}\right)\right] + o(k(\theta)).$$

Since ϕ is piecewise linear, the quotient under the sign of H_0 is equal to the derivative of ϕ on the interval from t to $t + \tau(\theta)$. Just as in the proof of Theorem 3.2.1, we obtain (3.2.4) for this class of functions, and from this class we pass to arbitrary functions; the only difference is that in the discrete case the abscissas of the vertices of the polygon l are chosen to be multiples of $\tau(\theta)$ (and so do depend on θ).

Now we proceed to prove (3.2.5). We use Theorem 2.3.2 with $i^{\theta} = k(\theta)(s - \Delta_1)$, $\delta' \le \delta/6$, $\Delta_1 = \gamma/8\,(T + 1 + s_0)$ instead of $\gamma/4\,(T + s_0)$ (see the proof of Theorem 3.2.2). The partition $\{t_m\}$ of the interval from 0 to T is chosen so that its points are multiples of $\tau(\theta)$ and so that $\Delta t_{\max} \le 2\Delta t_{\min}$. By repeating 3.2.3 and 3.2.4 we obtain the following estimate:

$$\mathsf{P}^{\theta}_{0, x_0}\{\rho_{0,T}(\xi^{\theta}, \tilde{\Phi}^{\theta}_{x_0; [0,T]}(k(\theta)(s - \Delta_1))) \ge 3\delta'\} \le$$

$$\le \exp\{-k(\theta)(s - \gamma)\}. \tag{3.2.8}$$

Here $\rho_{0,T}$ is the distance between functions considered only at multiples of $\tau(\theta)$.

The next stage is as follows. We extend the definition of an arbitrary function $\phi(t)$, $t = k\tau(\theta) \in [0, T]$, in the set $\tilde{\Phi}^{\theta}_{x_0; [0,T]}(k(\theta)(s - \Delta_1))$ by taking it linear between adjacent points $k\tau(\theta) \in [0, T]$; on the interval from $[T/\tau(\theta)]\,\tau(\theta)$ to T we make it constant. By Theorem 3.1.2 b) we obtain

$$\int_0^T H_0(t, \phi(t); \dot{\phi}(t))\, dt \le s,$$

i.e. $\phi \in \tilde{\Phi}^{\theta}_{x_0; [0,T]}(s)$.

The distance between ξ^θ and such a ϕ over the whole interval $[0, T]$ differs from their distance measured only over the multiples of $\tau(\theta)$ arbitrarily little, and from (3.2.8) we get (3.2.5). ◊

3.2.7. Remarks

Remark 1 (to Theorems 3.2.3, 3.2.3'). Conditions (3.2.1) - (3.2.3) are satisfied if $G^\theta(t, x; 0) = G_0(t, x; 0) = 0$; the function $G_0(t, x; z)$ has first and second derivatives in z, bounded when z changes within a bounded range;

$$\nabla_z (k(\theta)^{-1} G^\theta(t, x; k(\theta) z))|_{z=0} \to \nabla_z G_0(t, x; 0) \tag{3.2.9}$$

as $\theta \to$, uniformly with respect to t, x; and

$$\frac{\partial^2}{\partial z_i \partial z_j}(k(\theta)^{-1} G^\theta(t, x; k(\theta) z)) \to \frac{\partial^2}{\partial z_i \partial z_j} G_0(t, x; z) \tag{3.2.10}$$

as $\theta \to$, uniformly with respect to t, x and over every bounded set of values of z.

Remark 2. Replace Condition **C** by the following condition: for any bounded set $K \subset R^r$,

$$\Delta_K H_0(h, \delta') = \sup_{\substack{|t-s| \le h \\ |x-y| \le \delta',\ x \in K \\ H_0(t, x; u) < \infty}} \frac{H_0(s, y; u) - H_0(t, x; u)}{1 + H_0(t, x; u)} \to 0$$

as $h \downarrow 0$, $\delta' \downarrow 0$. In Conditions (3.2.1) - (3.2.3) we will not require uniformity over all $x \in R^r$ but only over every bounded set. Supplement the weakened requirement (3.2.1) by the following:

$$k(\theta)^{-1} G^\theta(t, x; k(\theta) z) \le \text{const} < \infty \tag{3.2.11}$$

for all t, all $x \in R^r$, all z in an arbitrary bounded set and all sufficiently far θ.

Then, under the conditions of Theorems 3.2.1, 3.2.3', $k(\theta) S_{0,T}(\phi)$ is the action functional for our family of processes, uniformly not over all initial points but only within every bounded set of values of x_0.

Remark 3. Let the stochastic processes under consideration and their characteristics depend on an additional parameter (next to θ), say a parameter α changing over a set A.

Let the conditions of Theorem 3.2.1 (those of Theorem 3.2.3', those mentioned in Remark 2) be fullfilled uniformly with respect to α. Then the statements of Theorem 3.2.3 (3.2.3', Remark 2) are fullfilled uniformly with respect to α.

3.3. Transition to manifolds. Actionfunctional theorems associated with truncated cumulants

3.3.1 Conditions A - E on a manifold

Let the manifold X satisfy Condition λ (for this condition as well as for notations involving manifolds, see 1.1.3). Let functions $G_0(t, x; z)$, $t \in [0, T]$, $x \in X$, $z \in T^*X_x$, and $H_0(t, x; u)$, $t \in [0, T]$, $x \in X$, $u \in TX_x$, be given, convex and associated by the Legendre transformation in the third argument. We impose the following restrictions on these functions:

A. There exist non-negative, downward convex functions

$$\overline{G}_0(z) < \infty, \; \underline{H}_0(u) \leq \infty \text{ of } z, u \in R^r,$$

associated by the Legendre transform, and such that for all $t \in [0, T]$, $x \in X$, $z \in T^*X_x$, $u \in TX_x$, and for any chart of the atlas considered whose domain contains a neighbourhood of x,

$$G_0(t, x; z) \leq \overline{G}_0(zA_x^{-1}), \quad H_0(t, x; u) \geq \underline{H}_0(A_x u).$$

B. If $\underline{H}_0(u)$ is finite, then so is $H_0(t, x; A_x^{-1}u)$ for any chart of our atlas whose domain contains a neighbourhood of x.

C. Define

$$\Delta H_0(h, \delta') = \sup \frac{H_0(t, y; A_y^{-1}u) - H_0(s, x; A_x^{-1}u)}{1 + H_0(s, x; A_x^{-1}u)},$$

where the supremum is taken over all t, s, $|t - s| \leq h$, $x, y \in X$, $\rho(x, y) \leq \delta'$, all charts of our atlas whose domain contains x and y, and all u, $\underline{H}_0(u) < \infty$. We require that $\Delta H_0(h, \delta') \to 0$ as $h \downarrow 0$, $\delta' \downarrow 0$.

D. The set $\{u: \underline{H}_0(u) < \infty\}$ is open, contains 0, and the least upper bound of the values of $H_0(t, x; 0)$ over all t, x is finite.

E. For any compactum $U_K \subset \{u: \underline{H}_0(u) < \infty\}$, the gradient $\nabla_u H_0(t, x; A_x^{-1})$ is bounded and continuous in $u \in U_K$ uniformly with respect to all t, x and all charts.

Let us note that Conditions **B** and **C**, especially in the case of $\{u: \underline{H}_0(u) < \infty\} \neq R^r$, are not invariant under change of the atlas by another atlas satisfying Condition λ also.

Define the functional $S_{T_1, T_2}(\phi)$ by formula (3.1.1); the notation $\Phi_{x; [T_1, T_2]}(s)$ has the same meaning as before.

Theorem 3.3.1. *Let the functions* $G_0 \leftrightarrow H_0$ *satisfy Conditions* **A** - **E**. *Then*

a) *the functions in the set* $\bigcup_{[T_1, T_2] \subseteq [0, T]} \bigcup_x \Phi_{x; [T_1, T_2]}(s)$ *are equicontinuous;*

b) *the functional* S_{T_1, T_2} *is lower semicontinuous with respect to uniform convergence;*

c) *for every compactum* $K \subseteq X$ *the set* $\bigcup_{x \in K} \Phi_{x; [T_1, T_2]}(s)$ *is compact.*

The **proof** of Theorem 3.1.1 remains valid in the case of a manifold. ◊

3.3.2. Formulation of Theorems 3.3.2 and 3.3.2'

Suppose that to each value of θ varying in a set Θ with a filter $\theta \rightarrow$ corresponds a locally infinitely divisible process $(\xi^\theta(t), \mathsf{P}^\theta_{t, x})$, $0 \le t \le T$, on X with compensating operator $\mathfrak{A}^\theta$ given by formula (1.3.1) with coefficients $b^{\theta, i}$, $a^{\theta, i, j}$ and measure $\lambda^\theta_{t, x}$. Let V be an open subset of $X \times X$, $V \supset \{(x, x): x \in X\}$, and $\rho(x, y) < \lambda/2$ if $(x, y) \in V$.

Define for each chart the truncated cumulant $G^\theta_V(t, x; z)$ of the process $(\xi^\theta(t), \mathsf{P}^\theta_{t, x})$ with respect to the set V (see § 2.4).

Theorem 3.3.2. *Let Conditions* **A** - **E** *be satisfied for the functions* G_0, H_0. *Let* $k(\theta)$ *be a real-valued function tending to* ∞ *as* $\theta \rightarrow$. *Let the following conditions be satisfied:*

$$\lim_{\theta \rightarrow} k(\theta)^{-1} \sup_{t, x} \ln \lambda^\theta_{t, x}\{y: (x, y) \notin V\} = -\infty; \tag{3.3.1}$$

$$k(\theta)^{-1} G^\theta_V(t, x; k(\theta) z) - G_0(t, x; zA_x) \rightarrow 0, \tag{3.3.2}$$

$$\nabla_z (k(\theta)^{-1} G^\theta_V(t, x; k(\theta) z) - G_0(t, x; zA_x)) \rightarrow 0 \tag{3.3.3}$$

as $\theta \rightarrow$, *uniformly with respect to* $t \in [0, T]$, *all charts* (W, ψ), *all* x *such that* $\rho(x, X \backslash W) > \lambda/4$, *and over every bounded set of values of* z. *For every bounded* $K \subset R^r$, *let*

$$\left| \frac{\partial^2}{\partial z_i \partial z_j} (k(\theta)^{-1} G^\theta_V(t, x; k(\theta) z)) \right| \le \text{const} < \infty \tag{3.3.4}$$

for all sufficiently far θ, *all* t, *all charts* (W, ψ), *all* x, $\rho(x, X \backslash W) > \lambda/4$, *and all* $z \in K$.

Then $k(\theta) S_{0, T}(\phi)$ *is the action functional for the family of processes* $\xi^\theta(t)$ *as* $\theta \rightarrow$, *uniformly with respect to the initial point* $x_0 \in X$.

We give the discrete version of this theorem (the generalization of Theorem 3.2.3').

Suppose to each $\theta \in \Theta$ corresponds a positive number $\tau(\theta)$ and a $\tau(\theta)$-process

$(\xi^\theta(t)\, \mathsf{P}^\theta_{t,x})$, $0 \le t \le T$, on X. Let V be a subset of $X \times X$ containing $\{(x, x): x \in X\}$, and let $\rho(x, y) < \lambda/2$ for $(x, y) \in V$; let $G^\theta_V(t, x; z)$ be the truncated cumulant of the process.

Theorem 3.3.2'. *Let Conditions* **A** - **E** *be satisfied for the functions* $G_0 \leftrightarrow H_0$; $k(\theta) \to \infty$, $\tau(\theta) \to 0$ *as* $\theta \to$. *Let*

$$\lim_{\theta \to} k(\theta)^{-1} \times \tag{3.3.1'}$$
$$\times \ln [\tau(\theta)^{-1} \sup_{t = k\tau(\theta),\, x \in X} \mathsf{P}^\theta_{t,x} \{(x, \xi^\theta(t + \tau(\theta))) \notin V\}] = -\infty$$

and let Conditions (3.3.2) - (3.3.4) *be satisfied.*

Then $k(\theta) S_{0,T}(\phi)$ *is the action functional for the family of processes* $\xi^\theta(t)$ *as* $\theta \to$, *uniformly with respect to the initial point.*

3.3.3. Proof of the theorems

We give the proof for the continuous case; for discrete time the proof differs only in that the times t_m (see below) are to be multiples of $\tau(\theta)$ (and so do depend on θ, which complicates the notations) and $\xi^\theta(t - \tau(\theta))$ is to be taken instead of $\xi^\theta(t-)$.

We have to demonstrate that for any positive δ, γ, s_0, for sufficiently far θ, for all $x_0 \in X$, all $\phi \in \Phi_{x_0;[0,T]}(s_0)$ and all $s \le s_0$,

$$\mathsf{P}^\theta_{0,x_0}\{\rho_{0,T}(\xi^\theta, \phi) < \delta\} \ge \exp\{-k(\theta)[S_{0,T}(\phi) + \gamma]\}, \tag{3.3.5}$$

$$\mathsf{P}^\theta_{0,x_0}\{\rho_{0,T}(\xi^\theta, \Phi_{x_0;[0,T]}(s)) \ge \delta\} \le \exp\{-k(\theta)(s - \gamma)\}. \tag{3.3.6}$$

Combined with the assertion of Theorem 3.3.1 this gives the assertion of our theorem.

Choose $h > 0$ in such a way that for functions ϕ with $S_{T_1,T_2}(\phi) \le s_0 + 2\gamma$ the inequality $|t - s| \le h$ implies $\rho(\phi(t), \phi(s)) < \lambda/4$. Choose a partition $0 = t_0 < t_1 < \dots < t_n = T$ of the interval from 0 to T with $\max_m (t_{m+1} - t_m) \le h$. Choose positive $\delta' < \delta$ and $\gamma' < \gamma$ (their choice will be made precise later). We verify that for sufficiently far θ, for all $x \in X$, all $\phi \in \Phi_{x;[t_m,t_{m+1}]}(s_0 + 2\gamma)$, and all $s \le s_0 + 2\gamma$,

$$\mathsf{P}^{\theta}_{t_m,\,x}\,\{\rho_{t_m,\,t_{m+1}}(\xi^{\theta},\,\phi) < \delta'\} \geq$$

$$\geq \exp\{-k(\theta)\,[S_{t_m,\,t_{m+1}}(\phi) + \gamma']\}, \tag{3.3.7}$$

$$\mathsf{P}^{\theta}_{t_m,\,x}\,\{\rho_{\{t_m,\,t_{m+1}\}}(\xi^{\theta},\,\Phi_{x;\,[t_m,\,t_{m+1}]}(s)) \geq \delta'\} \leq$$

$$\leq \exp\{-k(\theta)\,(s - \gamma')\}. \tag{3.3.8}$$

(We remind the reader of the notation $\rho_{\{t_m,\,t_{m+1}\}}$ introduced at the end of § 2.4: this is the usual distance for functions with coincident values at the ends of the interval and $+\infty$ otherwise).

This part of the proof is carried out as follows.

For a given initial point x we fix a chart (W, ψ) such that $\rho\,(x, X\backslash W) > \lambda$. Let us denote by B the $(\lambda/2)$-neighbourhood of the point x. Take

$$\tau^{\theta}_{B}(t_m) = \min\{t \geq t_m\colon \xi^{\theta}(t) \notin B\},$$

$$\tau^{\theta}_{V}(t_m) = \min\{t > t_m\colon (\xi^{\theta}(t-),\, \xi^{\theta}(t)) \notin V\}.$$

By Condition (3.3.1),

$$\mathsf{P}^{\theta}_{t_m,\,x}\,\{\tau^{\theta}_{V}(t_m) \leq t_{m+1}\} \leq (t_{m+1} - t_m)\,\sup_{t,\,y}\lambda^{\theta}_{t,\,y}\,\{y'\colon (y,\,y') \notin V\} \leq$$

$$\leq \exp\{-k(\theta)\cdot(s_0 + \gamma)\}.$$

Using Theorems 2.4.1, 2.3.2 with initial time t_m instead of 0 and reproducing the proof of Theorems 3.2.1, 3.2.2, we obtain that for sufficiently far θ, for all initial points x, all $\phi \in \Phi_{x;\,[t_m,\,t_{m+1}]}(s_0 + 2\gamma)$, and all $s \leq s_0 + 2\gamma$,

$$\mathsf{P}^{\theta}_{t_m,\,x}\,\{\rho_{t_m,\,t_{m+1}}(\xi^{\theta},\,\phi) < \delta'\} \geq \exp\{-k(\theta)\,[S_{t_m,\,t_{m+1}}(\phi) + \gamma'/2]\} -$$

$$- \mathsf{P}^{\theta}_{t_m,\,x}\,\{\tau^{\theta}_{B}(t_m) \leq t_{m+1}\}, \tag{3.3.9}$$

$$\mathsf{P}^{\theta}_{t_m,\,x}\,\{\rho_{t_m,\,t_{m+1}\wedge\tau^{\theta}_{B}(t_m)}(\xi^{\theta},\,\Phi_{x;\,[t_m,\,t_{m+1}\wedge\tau^{\theta}_{B}(t_m)]}(s)) \geq \delta'\} \leq$$

$$\leq \exp\{-k(\theta)\,(s - \gamma'/2)\}, \tag{3.3.10}$$

$$\mathsf{P}^{\theta}_{t_m, x}\{\rho_{[t_m, t_{m+1}]}(\xi^{\theta}, \Phi_{x; [t_m, t_{m+1}]}(s)) \geq \delta'\} \leq$$

$$\leq \exp\{-k(\theta)(s - \gamma/2)\} + \mathsf{P}^{\theta}_{t_m, x}\{\tau^{\theta}_B(t_m) \leq t_{m+1}\}. \tag{3.3.11}$$

Now, for sufficiently far θ,

$$\mathsf{P}^{\theta}_{t_m, x}\{\tau^{\theta}_B(t_m) \leq t_{m+1}\} \leq \exp\{-k(\theta)(s_0 + \gamma)\}.$$

Indeed, the functions in $\Phi_{x; [t_m, t_{m+1} \wedge \tau^{\theta}_B(t_m)]}(s_0 + 2\gamma)$ lie in the $(\lambda/4)$-neighbourhood of x, and their distance from the paths ξ^{θ} with $\tau^{\theta}_B(t_m) \leq t_{m+1}$ is greater than $\lambda/4$. Thus we obtain, by using (3.3.10), the estimate required.

This and formulas (3.3.9), (3.3.11) imply that for sufficiently far θ (3.3.7), (3.3.8) are fullfilled.

3.3.4 The estimate from below

Let us prove the estimate (3.3.5). For a sufficiently small positive $\delta'' \leq \delta$, for any function $\phi \in \bigcup_{x_0} \Phi_{x_0; [0, T]}(s)$, and for x from the δ''-neighbourhood of the point $\phi(t_m)$, the function

$$\phi_{m, x}(t) = \psi^{-1}(\psi(x) + \psi(\phi(t)) - \psi(\phi(t_m))), \quad t_m \leq t \leq t_{m+1},$$

is defined (where a map ψ whose domain W contains the λ-neighbourhood of the point x is used). This is the function whose image on the chart is obtained by shifting the image of $\phi(t)$ on the time interval from t_m to t_{m+1} by a vector such that $\phi_{m, x}(t_m)$ coincides with x. We impose one more restriction on the choice of δ'': we must have $\Delta H_0(0, \delta'') < \gamma/2n(1 + s_0)$. Then for all x such that $\rho(x, \phi(t_m)) < \delta''$,

$$S_{t_m, t_{m+1}}(\phi_{m, x}) \leq S_{t_m, t_{m+1}}(\phi) + \frac{\gamma}{2n}. \tag{3.3.12}$$

Now choose a positive $\delta' \leq \delta''$ so that for $x_0 = \phi(0)\ (= \phi(t_0))$ the inequalities

$$\rho(x_1, \phi_{0, x_0}(t_1)) < \delta',$$

$$\rho(x_2, \phi_{1, x_1}(t_2)) < \delta', \ \ldots, \ \rho(x_{n-1}, \phi_{n-2, x_{n-2}}(t_{n-1})) < \delta'$$

imply

$$\rho_{t_m, t_{m+1}}(\phi_{m, x_m}, \phi) < \delta'', \quad m = 0, 1, \ldots, n - 1$$

(δ' can be expressed in terms of δ'', n and the least upper and greatest lower bounds of the ratio $|\psi(y) - \psi(x)| / \rho(x, y)$). We then have

$$P^{\theta}_{0, x_0}\{\rho_{0,T}(\xi^{\theta}, \phi) < \delta\} \geq$$

$$\geq P^{\theta}_{0, x_0}\left(\bigcap_{m=0}^{n-1}\{\rho_{t_m, t_{m+1}}(\xi^{\theta}, \phi_{m, \xi^{\theta}(t_m)}) < \delta'\right\}.$$

(The figure helping to explain this formula should be like this: imagine a wide δ-tube about the function ϕ; insert a system of little tubes forming a telescope in it: the first section has initial radius δ', and every successive section is by δ' wider than the previous one.) Using the Markov property with respect to the times t_m, $m = 1, 2, ..., n - 1$, we obtain

$$P^{\theta}_{0, x_0}\{\rho_{0,T}(\xi^{\theta}, \phi) < \delta\} \geq$$

$$\geq \prod_{m=0}^{n-1} \inf_{\rho(x_m, \phi(t_m)) < \delta''} P^{\theta}_{t_m, x_m}\{\rho_{t_m, t_{m+1}}(\xi^{\theta}, \phi_{t_m, x_m}) < \delta'\}.$$

From this, using (3.3.7) with $\gamma' = \gamma/2n$ and inequality (3.3.12), we deduce (3.3.5).

3.3.5. The estimate from above

Now we prove (3.3.6). Introduce a distance $\rho_{\{t_0, t_1, ..., t_n\}}$ by putting

$$\rho_{\{t_0, t_1, ..., t_n\}}(\phi, \psi) = \sup_{0 \leq t \leq T} \rho(\phi(t), \psi(t)),$$

if

$$\phi(t_m) = \psi(t_m) \quad \text{for } m = 0, 1, ..., n,$$

and

$$\rho_{\{t_0, t_1, ..., t_n\}} = \infty$$

otherwise. Introduce random variables

$$\sigma_{0,T} = \min\{S_{0,T}(\phi): \phi(0) = \xi^{\theta}(0),\ \rho_{0,T}(\xi^{\theta}, \phi) \leq \delta\},$$

$$\sigma_{\{t_0, t_1, ..., t_n\}} = \min\{S_{0,T}(\phi): \rho_{\{t_0, t_1, ..., t_n\}}(\xi^{\theta}, \phi) \leq \delta'\},$$

$$\sigma_{\{t_m, t_{m+1}\}} = \min\{S_{t_m, t_{m+1}}(\phi): \rho_{\{t_m, t_{m+1}\}}(\xi^{\theta}, \phi) \leq \delta\}$$

(the minimum is attained by Theorem 3.3.1). For paths with initial point $\xi^\theta(0) = x_0$ the inequality $\rho_{0,T}(\xi^\theta, \Phi_{x_0;[0,T]}(s)) > \delta$ implies $\sigma_{0,T} > s$; hence it is sufficient to derive an estimate from above for $\mathsf{P}^\theta_{0,x_0}\{\sigma_{0,T} > s\}$. Besides, it is clear that $\sigma_{0,T} \le \sigma_{\{t_0, t_1, \ldots, t_n\}}$, and the latter random variable is equal to $\sum_{m=0}^{n-1} \sigma_{\{t_m, t_{m+1}\}}$ (by virtue of the equality $S_{0,T}(\phi) = \sum_{m=0}^{n-1} S_{t_m, t_{m+1}}(\phi)$).

So the following chain of inequalities holds:

$$\mathsf{P}^\theta_{0,x_0}\{\rho_{0,T}(\xi^\theta, \Phi_{x_0;[0,T]}(s)) > \delta\} \le \mathsf{P}^\theta_{0,x_0}\left\{\sum_{m=0}^{n-1} \sigma_{\{t_m, t_{m+1}\}} > s\right\} \le$$

$$\le \sum_{i_0=0}^{[s_0/\gamma]+1} \cdots \sum_{i_{n-2}=0}^{[s_0/\gamma]+1} \mathsf{P}^\theta_{0,x_0}\{(i_m - 1)\gamma' < \sigma_{\{t_m, t_{m+1}\}} \le i_m \gamma',$$

$m = 0, 1, \ldots, n-2;$

$$\sigma_{\{t_{n-1}, t_n\}} > s - (i_0 + i_1 + \ldots + i_{n-2})\gamma'\} + \sum_{m=0}^{n-2} \mathsf{P}^\theta_{0,x_0}\{\sigma_{\{t_m, t_{m+1}\}} > s_0\}.$$

Using the Markov property with respect to the times $t_1, \ldots, t_{n-1}$, we can continue this chain as follows:

$$\mathsf{P}^\theta_{0,x_0}\{\rho_{0,T}(\xi^\theta, \Phi_{x_0;[0,T]}(s)) > \delta\} \le$$

$$\le \sum_{i_0=0}^{[s_0/\gamma]+1} \cdots \sum_{i_{n-2}=0}^{[s_0/\gamma]+1} \prod_{m=0}^{n-2} \sup_x \mathsf{P}^\theta_{t_m, x}\{\sigma_{\{t_m, t_{m+1}\}} > (i_m - 1)\gamma\} \times$$

$$\times \sup_x \mathsf{P}^\theta_{t_{n-1}, x}\{\sigma_{\{t_{n-1}, t_n\}} > s - (i_0 + \ldots + i_{n-2})\gamma'\} +$$

$$+ \sum_{m=0}^{n-2} \sup_x \mathsf{P}^\theta_{t_m, x}\{\sigma_{\{t_m, t_{m+1}\}} > s_0\}.$$

Each factor in the large sum and each summand in the small sum is estimated by means of (3.3.8):

$$\mathsf{P}^{\theta}_{t_m, x}\{\sigma_{\{t_m, t_{m+1}\}} > (i_m - 1)\gamma'\} \leq$$
$$\leq \mathsf{P}^{\theta}_{t_m, x}\{\rho_{\{t_m, t_{m+1}\}}(\xi^{\theta}, \Phi_{x; [t_m, t_{m+1}]}((i_m - 1)\gamma')) > \delta'\} \leq$$
$$\leq \exp\{-k(\theta)(i_m - 2)\gamma'\},$$
$$\mathsf{P}^{\theta}_{t_{n-1}, x}\{\sigma_{\{t_{n-1}, t_n\}} > s - (i_0 + \ldots + i_{n-2})\gamma'\} \leq$$
$$\leq \exp\{-k(\theta)(s - (i_0 + \ldots + i_{n-2} + 1)\gamma')\},$$
$$\mathsf{P}^{\theta}_{t_m, x}\{\sigma_{\{t_m, t_{m+1}\}} > s_0\} \leq \exp\{-k(\theta)(s_0 - \gamma')\}.$$

Hence we obtain

$$\mathsf{P}^{\theta}_{0, x_0}\{\rho_{0, T}(\xi^{\theta}, \Phi_{x_0; [0, T]}(s)) > \delta\} \leq$$
$$\leq ([s_0/\gamma'] + 2)^{n-1} \exp\{-k(\theta)(s - (2n - 1)\gamma')\} +$$
$$+ (n - 1)\exp\{-k(\theta)(s_0 - \gamma')\}.$$

Taking $\gamma' = \gamma/2n$ we obtain (3.3.6). ◊

CHAPTER 4

SPECIAL CASES

4.1. Conditions A - E of § 3.1 - § 3.3

Many results in this chapter will be formulated using the requirements that the functions $G_0(t, x; z) \leftrightarrow H_0(t, x; u)$ involved satisfy Conditions **A** - **E**. In this section we give some examples verifying the fulfillment of these conditions.

4.1.1. Quadratic functions $G_0 \leftrightarrow H_0$

First of all, consider the case

$$G_0(t, x; z) = \sum_i b^i(t, x)\, z_i + \frac{1}{2}\sum_{i,j} a^{ij}(t, x)\, z_i z_j \tag{4.1.1}$$

(which corresponds, in the discrete case, to processes with normally distributed jumps, and in the continuous case, to diffusion processes). If the matrix $(a^{ij}(t, x))$ is non-degenerate, the Legendre transform is given by

$$H_0(t, x; u) = \frac{1}{2}\sum_{i,j} a_{ij}(t, x)\,(u^i - b^i(t, x))\,(u^j - b^j(t, x)), \tag{4.1.2}$$

where $(a_{ij}) = (a^{ij})^{-1}$. It is easily checked that if the coefficients $b^i(t, x)$, $a^{ij}(t, x)$ are bounded and uniformly continuous while the matrix $(a^{ij}(t, x))$ is uniformly positive definite, Conditions **A** - **E** are satisfied. In particular, the functions $\overline{G}_0 \leftrightarrow \underline{H}_0$ are given by

$$\overline{G}_0(z) = B\,|z| + \frac{1}{2}A\,|z|^2;$$

$$\underline{H}_0(u) = \begin{cases} 0, & |u| \le B, \\ \frac{1}{2}A^{-1}(|u| - B)^2, & |u| > B, \end{cases}$$

where B is the least upper bound of $|b(t, x)|$ and A is the least upper bound of the eigenvalues of the matrix $(a^{ij}(t, x))$.

This remains true for manifolds satisfying Condition λ.

4.1.2. A class of examples on a Riemannian manifold

Let X be a compact three-dimensional Riemannian manifold of class $C^{(2)}$. For vectors of the tangent and the cotangent spaces TX_x, T^*X_x, we denote by $|\ |_x$ the Riemannian length of a vector. Let the function $G_0(t, x; z) \equiv G_0(x; z)$ be given by

$$G_0(x; z) = \ln \int_{TX_x} e^{zu}\, \mu_x(du),$$

where μ_x, for each x, is a probability distribution on TX_x, uniform over all spheres and such that $\mu_x\{u: |u|_x \in d\rho\} = \mu(d\rho)$, where the radial distribution μ is some distribution on the half-line $[0, \infty)$, one and the same for all points x. Such a function G_0 appears, in particular, for one of the discrete examples considered in 4.3.5.

Calculation leads to $G_0(x; z) = F_0(|z|_x)$, where

$$F_0(t) = \ln \int_0^\infty \frac{\operatorname{sh} \rho t}{\rho t}\, \mu(d\rho).$$

The Legendre transform also depends only on the length of a vector $u \in TX_x$: $H_0(x; u) = A_0(|u|_x)$, where $A_0 \leftrightarrow F_0$. If $\ln \mu[\rho, \infty) \sim -K\rho^\beta$ as $\rho \to \infty$, where $K > 0$ and β is some number greater than 1, then Conditions **A** - **E** are satisfied by the functions G_0, H_0.

First of all, an atlas satisfying Condition λ does exist and there even exists, by the compactness of X, such an atlas consisting of a finite number of charts; choose such an atlas. For the function $\overline{G}_0(z)$ we take $F_0(c|z|)$, where c is a sufficiently large positive constant; $\underline{H}_0(u) = A_0(c^{-1}|u|)$. Hence we obtain that Condition **A** is satisfied. Next we verify that $F_0(t) \sim K' t^{\frac{\beta}{\beta-1}}$ as $t \to \infty$, which implies that A_0 is finite for all values of its argument (and thus Conditions **B** and **D** hold). Condition **E** is checked easily. We verify that **C** is fulfilled.

An equivalent form of this condition is that for any $\varepsilon \in (0, 1)$ there exists a $\delta > 0$ such that for any x, y in a single chart, $\rho(x, y) < \delta$, and for any $z \in R^3$ we have $G_0(y; (1-\varepsilon) z A_y) \le \varepsilon + (1-\varepsilon) G_0(x; z A_x)$, i.e.

$$F_0((1-\varepsilon)|zA_y|_y) \le \varepsilon + (1-\varepsilon) F_0(|zA_x|_x). \tag{4.1.3}$$

For sufficiently small $\rho(x, y)$ we have

$$|zA_y|_y \le (1-\varepsilon)^{-\frac{1}{3\beta}} |zA_x|_x .$$

At the same time, for sufficiently large t,

$$K' (1-\varepsilon)^{\frac{1}{3\beta}} t^{\frac{\beta}{\beta-1}} < F_0(t) < K' (1-\varepsilon)^{-\frac{1}{3\beta}} t^{\frac{\beta}{\beta-1}}.$$

Therefore, for sufficiently large $|z|$ and small $\rho(x, y)$,

$$F_0((1-\varepsilon)\,|z A_y|_y) <$$

$$< K' (1-\varepsilon)^{-\frac{1}{3\beta}} [(1-\varepsilon)^{1-\frac{1}{3\beta}} |z A_x|_x]^{\frac{\beta}{\beta-1}} <$$

$$< K' (1-\varepsilon)^{\frac{3\beta-1}{3\beta-3}} |z A_z|_x^{\frac{\beta}{\beta-1}} < (1-\varepsilon) F_0(|z A_x|_x).$$

This implies (4.1.3) for large $|z|$, say, for $|z| > z_0$.

As for the remaining z, it is easily seen that the function $G_0(x; z A_x)$ is continuous with respect to x, z, uniformly with respect to all $z \in R^3$, $|z| \le z_0$, all charts and all x in the domain of a given chart. So (4.1.3) is fulfilled for all values of z.

4.1.3. Theorem 4.1.3

In this example we restrict ourselves to the one-dimensional case. Let the function G_0 be given by the formula

$$G_0(t, x; z) = b(t, x)\, z + \int_{-\infty}^{\infty} (e^{zu} - 1 - zu)\, u^{-2} \mu_{t,x}(du),$$

where the function $(e^{zu} - 1 - zu)\, u^{-2}$ is defined by continuity at $u = 0$ (the function G_0 will have such a form in the continuous examples considered in § 4.3).

Theorem 4.1.1. *Let $b(t, x)$ be a bounded, uniformly continuous function, let the measure $\mu_{t,x}$ be bounded, uniformly weakly continuous with respect to t, x, and, for every t, x, be concentrated on an interval $[u_-(t, x),\ u_+(t, x)]$, where the functions $u_-(t, x)$, $u_+(t, x)$ are bounded and uniformly continuous, $u_-(t, x) < -c$, $u_+(t, x) > c$, $c > 0$. Let, for every $\kappa > 0$, there exist a $\lambda > 0$ such that*

$$\mu_{t,x}[u_-(t, x),\ u_-(t, x)(1-\kappa)] > \lambda,$$

$$\mu_{t,x}[(1-\kappa)\, u_+(t, x),\ u_+(t, x)] > \lambda$$

for all t, x. Then Conditions **A** - **E** *hold for the functions $H_0 \leftrightarrow G_0$.*

Proof. First of all, it is easily seen that $\overline{G}_0$ can be taken to be the function $\overline{G}_0(z) = C\,(e^{A|z|} - 1)$, where A and C are some positive constants; $\underline{H}_0(u) = Ch(|u|/AC)$, where

$$h(v) = \begin{cases} v \ln v - v + 1 & \text{for } v \geq 1, \\ 0 & \text{for } 0 \leq v < 1. \end{cases}$$

We verify Condition **C**: for any $\varepsilon \in (0, 1)$, for sufficiently close (t, x), (s, y), and for all z,

$$G_0(t, x; (1-\varepsilon) z) - (1-\varepsilon) G_0(s, y; z) \leq \varepsilon. \tag{4.1.4}$$

For a fixed ε take $\kappa = \varepsilon/3$; choose positive h and δ', to begin with, in such a way that $u_-(t, x) / u_-(s, y)$, $u_+(t, x) / u_+(s, y)$ exceed $1 - \kappa$ for $|t - s| \leq h$, $|x - y| \leq \delta'$. For $z > 0$ we have

$$G_0(t, x; (1-\varepsilon) z) \leq$$

$$\leq Bz + \frac{\mu_{t,x}(R^1)\left(e^{(1-\varepsilon) u_+(t,x) z} - 1 - (1-\varepsilon) u_+(t, x) z\right)}{u_+(t, x)^2} \tag{4.1.5}$$

(here $B = \sup |b(t, x)|$); recall that $\mu_{t,x}(R^1) \leq M = \text{const} < \infty$. In turn,

$$(1-\varepsilon) G_0(s, y; z) \geq$$

$$\geq \lambda \frac{e^{(1-\kappa) u_+(s,y) z} - 1 - (1-\kappa) u_+(s, y) z}{[(1-\kappa) u_+(s, y)]^2}(1-\varepsilon) - Bz. \tag{4.1.6}$$

For sufficiently large z ($z > z_0$) expression (4.1.5) is less than (4.1.6) if only $|t-s| \leq h$, $|x - y| \leq \delta'$; that is, for $z > z_0$ the left-hand side of (4.1.4) is non-positive. The same holds for negative z that are sufficiently large in absolute value, i.e. for all z, $|z| > z_0$. For $|z| \leq z_0$ the function $G_0(t, x; z)$ is uniformly continuous with respect to all arguments; therefore we can diminish h, δ' in such a way that

$$|G_0(t, x; (1-\varepsilon) z) - G_0(s, y; (1-\varepsilon) z)| < \varepsilon$$

for all $|t-s| \leq h$, $|x-y| \leq \delta'$, $|z| \leq z_0$. It is easy to prove that

$$(e^{(1-\varepsilon) zu} - 1 - (1-\varepsilon) zu) u^{-2} \leq (1-\varepsilon)(e^{zu} - 1 - zu) u^{-2}$$

for all u, z, from which

$$G_0(s, y; (1-\varepsilon) z) \leq (1-\varepsilon) G_0(s, y; z)$$

and, finally,

$$G_0(t, x; (1-\varepsilon) z) - (1-\varepsilon) G_0(s, y; z) =$$
$$= G_0(t, x; (1-\varepsilon) z) - G_0(s, y; (1-\varepsilon) z) +$$
$$+ G_0(s, y; (1-\varepsilon) z) - (1-\varepsilon) G_0(s, y; z) < \varepsilon.$$

The verification of the remaining conditions is easy. ◊

4.1.4. A class of examples related to finitely many vectors

Let X be a manifold satisfying Condition λ; let to every $t \in [0, T]$ and to every point $x \in X$ correspond n vectors $u_1(t, x), \ldots, u_n(t, x)$ of TX_x and n positive numbers $\lambda_1(t, x), \ldots, \lambda_n(t, x)$. Define the function G_0 by

$$G_0(t, x; z) = \sum_{i=1}^{n} \lambda_i(t, x)\,(e^{zu_i(t, x)} - 1).$$

Let all functions $u_i(t, x)$, $\lambda_i(t, x)$ be uniformly continuous and bounded (for the functions u_i this means that the $A_x u_i(t, x)$ are bounded for any chart of our atlas satisfying Condition λ); suppose $\lambda_i(t, x) \geq \lambda_0 > 0$. Denote by $U_{t, x} \subset TX_x$ the convex hull of the set $\{u_1(t, x), \ldots, u_n(t, x)\}$. Let the greatest lower bound of $|A_x u|$ over $u \notin U_{t, x}$, over all t, x, and over all charts be positive. Then for the function G_0 and its Legendre transform H_0 Conditions **A** - **E** are satisfied.

The proof is the same as in the previous subsection; the condition on the convex hull plays the rôle of the condition $u_-(t, x) < -c$, $u_+(t, x) > c$, $c > 0$, and the uniform positivity of $\lambda_i(t, x)$ provides the uniform "sufficient filling" of the extremities of the set $U_{t, x}$ by the measure.

4.2. Patterns of processes with frequent small jumps. The cases of very large deviations, not very large deviations, and super-large deviations

4.2.1. Very large, not very large and super-large deviations in the case of discrete time

We will consider families of Markov processes depending on two positive parameters τ and h; roughly speaking, the time between the jumps will be proportional to τ, and their sizes proportional to h.

First we consider the case of processes in R^r with discrete time.

Suppose that for all $t \in [0, T)$, $x \in R^r$ we are given probability distributions $\mu_{t, x}$ on R^r; for $\tau > 0$, $h > 0$ let a vector function $b^{\tau, h}(t, x)$ of $t \in [0, T)$, $x \in R^r$, be given. Suppose these functions depend on x in a measurable way.

Consider for each $\tau > 0$, $h > 0$, a τ-process $(\xi^{\tau,h}(t), \mathsf{P}^{\tau,h}_{t,x})$, $0 \le t \le T$, in R^r such that if the process is at a point x at a time $t = k\tau$ $(k = 0, 1, 2, \dots)$, it performs a jump of size $\tau \cdot b^{\tau,h}(t, x) + h \cdot U$ at time $t + \tau$, where the random variable U has distribution $\mu_{t,x}$. In other words, if t is a multiple of τ,

$$\mathsf{P}^{\tau,h}_{t,x}\ \{\xi^{\tau,h}(t+\tau) \in \Gamma\} = \mu_{t,x}(h^{-1}(\Gamma - x - \tau b^{\tau,h}(t,x))).$$

If for all t, x the whole distribution $\mu_{t,x}$ is concentrated at the point 0 and $b^{\tau,h}(t, x) \equiv b(t, x)$, the paths of $\xi^{\tau,h}(t)$ are almost the same as the Euler polygons for the differential equation $\dot{x}(t) = b(t, x(t))$. For a non-trivial $\mu_{t,x}$ our stochastic process can be considered as the result of a "noise" perturbing this differential equation; however, we must bear in mind that these perturbations may be larger in size and may play a rôle not less significant than the "unperturbed" motion itself.

We will consider asymptotic problems for the processes $(\xi^{\tau,h}(t), \mathsf{P}^{\tau,h}_{t,x})$ as $\tau \to 0$, $h \to 0$. Such problems, however, do not necessarily involve large deviations for every relation between the rates of convergence of τ and h to zero. To make this clear, consider the corresponding problems for sums of n identically distributed, independent random variables; they are covered by our pattern with

$$b^{\tau,h}(t, x) \equiv 0,\ \ \mu_{t,x} \equiv \mu,\ \ \tau = 1/n.$$

In this case the process $\xi^{\tau,h}(t)$ starting at the point 0 at time 0 can be represented as $h(X_1 + X_2 + \dots + X_{[nt]})$, where $X_1, X_2, \dots$ are independent random variables with distribution μ. Suppose the expectation $\int u\,\mu\,(du) = 0$ and the variance $\int u^2 \mu\,(du) = 1$ (we consider the one-dimensional case). The probability $\mathsf{P}^{\tau,h}_{0,0}\{\xi^{\tau,h}(1) > 1\}$ is the same as $\mathsf{P}\ \{(X_1 + \dots X_n)/\sqrt{n} > \tau^{1/2} h^{-1}\}$ (we write the probability sign without subscripts or superscripts). If $\tau^{1/2} h^{-1}$ does not tend to infinity, then the problem of finding the asymptotics of this probability is not a problem on large deviations at all. We will consider only cases in which $\tau \to 0$, $\tau^{1/2} h^{-1} \to \infty$ ($h \to 0$ follows from these conditions). In such cases it is, however, necessary to introduce a classification of problems on large deviations.

We will call problems concerning deviations from 0 of order $\sqrt{n}$ of a normalized sum of n independent random variable problems on *very large deviations*; those concerning deviations of order $o(\sqrt{n})$ (but tending to infinity) will be called problems on *not very large deviations*; deviations going to infinity faster than $\sqrt{n}$ are called *super-large*. As is known, under Cramér's condition of finiteness of exponential moments, if the size of deviations goes from not very large via very large to super-large, the character of results on large deviations undergoes a qualitative change. The results on not very large deviations are still related to the normal distribution (see, for example, Theorem 1 of Cramér [1]; of course, one must keep in mind that this is a precise result and not a rough one, that is, one up to logarithmic equivalence, as is the case in these chapters). The results on very large deviations (Theorem 6 of Cramér [1]) are not related to the normal distribution and they are formulated in terms of Legendre transforms. Results on super-large deviations have been obtained in some papers (see, e.g., Dzhakhangirova, Nagaev [1]); they depend very strongly on the behaviour of the "tails" of the distribution μ.

In accordance with this, within our general pattern we will speak about the case $\tau = h$ (or τ and h of the same order) as the case of *very large deviations*; the case of $\tau = o(h)$ (one must not forget that at the same time $\tau^{1/2} h^{-1} \to \infty$, i.e., $h = o(\tau^{1/2})$), as the case of *not very large deviations*; and that of $h = o(\tau)$, $\tau \to 0$, as the case of *super-large deviations*. In the case of not very large deviations we will require that $\int u\mu_{t,x}(du) = 0$ (if this condition is not imposed, in the case of normalized sums of independent random variables the expectation will dwarf the deviation of order $o(\sqrt{n})$); in the cases of very large and super-large deviations this requirement is not essential. In all cases we will suppose that $b^{\tau,h}(t, x) \to b(t, x)$.

4.2.2. The cumulant

The cumulant $G^{\tau,h}$ of the process $\xi^{\tau,h}(t)$ is given by

$$G^{\tau,h}(t, x; z) = \tau^{-1} \ln \mathsf{M}^{\tau,h}_{t,x} \exp\{z(\xi^{\tau,h}(t+\tau) - x)\} =$$

$$= zb^{\tau,h}(t, x) + \tau^{-1} G_*(t, x; hz), \tag{4.2.1}$$

$$G_*(t, x; z) = \ln \int_{R^r} e^{zn} \mu_{t,x}(du). \tag{4.2.2}$$

The results on not very large deviations will be obtained by using the MacLaurin expansion of $G_*(t, x; z)$ in z.

4.2.3. Continuous time

In the continuous case we will consider families of locally infinitely divisible processes $(\xi^{\tau, h}(t), \mathsf{P}^{\tau, h}_{t, x})$, $0 \le t \le T$, in R^r, depending on parameters τ, $h > 0$, with compensating operators of the form

$$\mathfrak{A}^{\tau, h} f(t, x) = \frac{\partial f}{\partial t}(t, x) + \sum_i b^{\tau, h, i}(t, x) \frac{\partial f}{\partial x^i}(t, x) +$$

$$+ \frac{\tau^{-1} h^2}{2} \sum_{i, j} a^{ij}(t, x) \frac{\partial^2 f}{\partial x^i \partial x^j}(t, x) +$$

$$+ \tau^{-1} \int_{R^r} \left[f(t, x + hu) - f(t, x) - h \sum_i \frac{\partial f}{\partial x^i}(t, x) u^i \right] \nu_{t, x}(du) \qquad (4.2.3)$$

and generators

$$A_t^{\tau, h} f(t, x) = \sum_i b^{\tau, h, i}(t, x) \frac{\partial f}{\partial x^i}(x) + \frac{\tau^{-1} h^2}{2} \sum_{i, j} a^{ij}(t, x) \frac{\partial^2 f}{\partial x^i \partial x^j}(x) +$$

$$+ \tau^{-1} \int_{R^r} \left[f(x + hu) - f(x) - h \sum_i \frac{\partial f}{\partial x^i}(x) u^i \right] \nu_{t, x}(du). \qquad (4.2.4)$$

The factor $\tau^{-1} h^2$ in front of the diffusion part is chosen here to match the jump part (i.e. the integral).

The case $\tau = h \to 0$ will be called the case of very large deviations; the case of $\tau h^{-1} \to 0$, $\tau h^{-2} \to \infty$ (whence $\tau \to 0$, $h \to 0$), that of not very large deviations; the case of $h \to 0$, $\tau h^{-1} \to \infty$, the case of super-large deviations (note that in the last case, in contrast to the discrete case, one need not require that τ tends to 0). In all cases we will assume that $b^{\tau, h}(t, x) \to b(t, x)$ as τ and h change in the way prescribed.

The cumulant $G^{\tau, h}$ in the continuous case can be expressed, as in the discrete case, by the formula

$$G^{\tau, h}(t, x; z) = z b^{\tau, h}(t, x) + \tau^{-1} G_*(t, x; hz), \qquad (4.2.5)$$

$$G_*(t, x; z) = \frac{1}{2} \sum_{i, j} a^{ij}(t, x) z_i z_j + \int_{R^r} (e^{zu} - 1 - zu) \nu_{t, x}(du). \qquad (4.2.6)$$

4.2.4. Processes with independent increments

An important particular case of these patterns is that of families of processes with independent increments, both for continuous and discrete time (i.e., processes of successive sums of independent random variables). This case is characterized by $\mu_{t,x} \equiv \mu_t$, $b^{\tau,h}(t, x) \equiv 0$ (introducing $b^{\tau,h}(t)$ not depending on x does not yield anything new). The theorems in the subsequent sections can be applied to the case of processes with independent increments. Large deviations for families of processes with independent increments were studied by Borovkov [1], Mogul'skii [1]; since the local characteristics do not depend on x, the results were obtained under weaker assumptions. Borovkov, Mogul'skii [1] obtained results on large deviations for processes with independent increments in infinite-dimensional spaces.

4.2.5. Examples with $\tau = h$

The case most often arising in applications is that of our pattern with $\tau = h$ (very large deviations), $b^{h,h}(t, x) \equiv b(t, x)$.

Let us consider an example. Suppose that in some volume V filled with nutritive medium bacteria are multiplying. Suppose after an infinitely small time interval dt each bacterium splits in halves with probability $\lambda_+ dt$ and dies with probability $\lambda_- dt$ (we postulate the Markov character of the process, which, of course, holds in a first approximation only). Let the birth and death rates λ_+ and λ_- depend on the availability of nutritive substance to the bacteria and, finally, on the concentration of bacteria in the given volume:

$$\lambda_+ = \lambda_+ (N(t)/V), \quad \lambda_- = \lambda_- (N(t)/V),$$

where $N(t)$ is the number of bacteria at time t.

We introduce the notation $\xi^V(t) = N(t)/V$. The process $\xi^V(t)$ depends on the parameter in the way described in 4.2.3, with $\tau = h = V^{-1}$. Indeed, the process ξ^V starting from a point x at time t is constant for an exponential time interval; in a time interval of length dt it jumps to the point $x + V^{-1}$ with probability $N(t)\lambda_+ dt = V \cdot x\lambda_+(x)\, dt$ and to the point $x - V^{-1}$ with probability $V \cdot x\lambda_-(x)\, dt$. The generator of the process is given by

$$A_t^V f(x) = V\,[x\lambda_+(x)\,(f(x + V^{-1}) - f(x)) + x\lambda_-(x)\,(f(x - V^{-1}) - f(x))].$$

It is the same as formula (4.2.4) with $\tau = h = V^{-1}$, $b(t, x) = x(\lambda_+(x) - \lambda_-(x))$ and measure $\nu_{t,x}$ concentrated at the points $+1$ and -1:

$$\nu_{t,x}\{1\} = x\lambda_+(x), \quad \nu_{t,x}\{-1\} = x\lambda_-(x).$$

In this example limit theorems for $\tau = h \to 0$ concern the asymptotic behaviour of the population of bacteria as the volume V tends to ∞ (that is, in essence, concerning the behaviour of a population of a large number of bacteria).

As a second example we will consider the empirical distribution function $F_n^*(t)$ constructed for n independent, identically distributed random variables $X_1, \ldots, X_n$. This random function is Markovian; it increases by jumps of size $1/n$, and the probability that a jump occurs on a time interval from t to $t + dt$ under the condition that $F_n^*(t) = x$ (assuming that the distribution of the random variables X_i is uniform over the interval (0, 1)) is equal to $n \frac{1-x}{1-t} dt$ $(0 \le t < 1,\ 0 \le x \le 1)$. The generator A_t^n is given by the formula

$$A_t^n f(x) = n \frac{1-x}{1-t} \left[f\left(x + \frac{1}{n}\right) - f(x) \right],$$

and it can be rewritten in the form (4.2.4) with $\tau = h = 1/n$.

4.2.6. Shifts along geodesics

Let X be a Riemannian manifold of class $C^{(2)}$. For $x \in X$, $u \in TX_x$, we denote by $e(x, u)$ the point of the manifold obtained by shifting x along the geodesic in the direction of u over a distance equal to the Riemannian length $|u|_x$ of the vector u (we suppose that each geodesic can be extended indefinitely). Let to each $t \in [0, T)$, $x \in X$, correspond a distribution $\mu_{t,x}$ on TX_x. Let for any $\tau > 0$, $h > 0$, a function $b^{\tau, h}(t, x)$ with values in TX_x be defined (the dependence on x is taken to be measurable).

Consider for $\tau > 0$, $h > 0$, the Markov stochastic process $(\xi^{\tau, h}(t), \mathsf{P}_{t,x}^{\tau, h})$ constructed as follows. If the process at time t (a multiple of τ) is at the point x, it stays there during the time interval $[t, t + \tau)$ and at time $t + \tau$ jumps to the point $e(x, \tau b^{\tau, h}(t, x) + hU)$, where U is a random vector of TX_x with distribution $\mu_{t,x}$.

Here the classification of large deviations remains the same.

Azencott, Ruget [1] obtained a theorem on large deviations for such a pattern in the case of very large deviations. In § 4.3 using the general results of the previous chapter, we will obtain an analogous result, but under somewhat different assumptions.

We will not introduce the continuous variant of this pattern on a manifold, because we do not intend to give concrete results of this type.

4.2.7. Relation with the normal distribution

The theorems on large deviations for stochastic processes we will obtain, as well as those for sums of independent random variables, retain some relation to the normal distribution and diffusion processes in the case of not very large deviations (see § 4.4, § 4.5); namely, the normalized action functional is the integral of a quadratic function H_0 of the form (4.1.2). In the case of very large deviations (§ 4.3) this relation is lost, the function H_0 can be expressed in terms of the exponential moments by means of the Legendre transform. For super-large deviations only very special results will be obtained.

4.2.8. The pattern in Varadhan [1]

One can consider various other patterns of Markov processes with small jumps. In particular, Varadhan [1] considers a pattern of sums of independent random variables with distributions chosen in a special way. We set forth the construction of this paper in our notations.

Let $g(z)$ be a bounded, downward convex function on the real line, $g(0) = 0$. Define the function $h(u)$ as its Legendre transform; it is supposed that $h(u) < +\infty$ in some interval. Let a_n be a sequence such that $a_n/n \to \infty$ $(n \to \infty)$. Consider independent random variables $X_{n,j}$ with density

$$c_n \exp\left\{-\frac{a_n}{n} h(u)\right\};$$

define

$$\xi_n(t) = \frac{1}{n}\sum_{j=1}^{[nt]} X_{n,j}.$$

Varadhan [1] proves that the action functional for the family of processes ξ_n as $n \to \infty$ is of the form $a_n S_{0,T}(\phi)$, where

$$S_{0,T}(\phi) = \int_0^T h(\dot{\phi}(t))\,dt.$$

This result does not fall into our classification of very large, not very large and super-large deviations.

Let us show how it can be deduced from Theorem 3.2.3' under the additional assumption that the functions $g(z)$, $h(u)$ are twice continuously differentiable (the latter, in the interval on which it is bounded).

The fulfilment of Conditions **A** - **E** is obvious. In this case n takes the rôle of parameter θ, and $\theta \to$ means $n \to \infty$; $k(\theta) = a_n$, $\tau(\theta) = n^{-1}$. Evaluate the cumulant corresponding to the parameter value n:

$$G_n(z) = n \ln \int \exp\left\{\frac{zu}{n}\right\} \cdot c_n \exp\left\{-\frac{a_n}{n} h(u)\right\} du.$$

Lemma 2.2 of Varadhan's paper states that as $n \to \infty$,

$$G_n(a_n z) = a_n [g(z) + o(1)]. \tag{4.2.7}$$

It is easy to make this statement more precise, proving that (4.2.7) holds uniformly with respect to z in any bounded region; this gives Condition (3.2.1). Applying Laplace's method to the integrals

$$\int u \exp\left\{\frac{a_n}{n} [zu - h(u)]\right\} du,$$

$$\int (u - u_0)^2 \exp\left\{\frac{a_n}{n} [zu - h(u)]\right\} du$$

gives Conditions (3.2.2), (3.2.3) too; using Theorem 3.2.3' we obtain the result required.

4.3. The case of very large deviations

4.3.1. Theorem 4.3.1. *Let a family of Markov processes of one of the two classes described in Subsections* 4.2.1, 4.2.3 *be given, with* $\tau = h$: $(\xi^{h,h}(t), \mathrm{P}^{h,h}_{t,x})$. *Let* $b^{h,h}(t, x) \to b(t, x)$ *as* $h \to 0$, *uniformly with respect to* t, x. *Let the functions* $G_0(t, x; z) = zb(t, x) + G_*(t, x; z)$ *and* $H_0(t, x; u) \leftrightarrow G_0(t, x; z)$ *satisfy Conditions* **A** - **E** *of* § 3.1 *and* § 3.2, *and let the functional* $S_{0,T}(\phi)$ *be given by* (3.1.1).

Then $h^{-1} S_{0,T}(\phi)$ *is the action functional for* $(\xi^{h,h}(t), \mathrm{P}^{h,h}_{t,x})$ *as* $h \to 0$, *uniformly with respect to the initial point.*

The **proof** is an immediate application of Theorems 3.2.3, 3.2.3'. It follows from the representations (4.2.2), (4.2.6) for the function $G_*(t, x; z)$ that this function (and therefore $G_0(t, x; z)$ too) is infinitely differentiable in z, and that for z running over any compact set the derivatives can be estimated, uniformly with respect to t, x, using the least upper bound of the function G_* itself on a larger set. Hence we obtain Conditions (3.2.2), (3.2.3) (the fact that (3.2.1) is fulfilled is trivial). ◊

4.3.2. Theorem 4.3.2. *Let* $(\xi^{h,h}(t), \mathsf{P}^{h,h}_{t,x})$, $h > 0$, *be a family of processes as in the previous theorem; let* $b^{h,h}(t, x) \equiv b(t, x)$. *Let the functions* $G_0(t, x; z) = zb(t, x) + G_*(t, x; z)$ *and* $H_0 \leftrightarrow G_0$ *satisfy* **A**, **B** *and* **C**, *and the following conditions instead of* **D**, **E**:

D'. *The set* $\{u: \underline{H}_0(u) < \infty\}$ *has at least one interior point* u_0, *and*

$$\sup_{t,x} H_0(t, x; u_0) < \infty.$$

D". *The set of points u of the closure* $\overline{U}$ *of the set* $\{u: \underline{H}_0(u) < \infty\}$ *for which* $\underline{H}_0(u) = \infty$ *is closed.*

E'. *For any compactum* $U_K \subseteq \{u: \underline{H}_0(u) < \infty\}$ *the function* $H_0(t, x; u)$ *is continuous in u uniformly with respect to t, x and* $u \in U_K$.

E". *For any compactum* U_K *consisting entirely of interior points of* $\{u: \underline{H}_0(u) < \infty\}$ *the gradient* $\nabla_u H_0(t, x; u)$ *is bounded and continuous with respect to u uniformly with respect to t, x,* $u \in U_K$.

Then $h^{-1}S_{0,T}(\phi)$ *is the action functional for the family of processes* $\xi^{h,h}(t)$ *as* $h \to 0$, *uniformly with respect to the initial point.*

The **proof** is a simplification of that of the theorems of § 3.2 (the simplification consists in that instead of Theorem 2.3.2 one uses Theorem 2.3.1, and there is no need for introducing a functional $\tilde{I}^h_{T_1, T_2}(\phi)$ different from the action functional); the remark to Theorems 2.3.1 and 2.3.2 is used, which enables us to restrict ourselves to the values of u belonging to the set $\overline{U}$. ◊

Theorem 4.3.2 admits a wider class of cumulants $G_0(t, x; z)$ than Theorem 4.3.1; in the one-dimensional case this class can be still widened. First, Condition **D"** is satisfied automatically because the set in question consists of two points at most; secondly, Condition **E'** can be replaced by the condition that the function $H_0(t, x; u)$ should be uniformly bounded on each compactum $U_K \subseteq \{u: \underline{H}_0(u) < \infty\}$.

4.3.3. A concrete example

Let functions $\lambda_1(t, x)$, $\lambda_{-1}(t, x)$, $\lambda_2(t, x)$ of $t \in [0, T)$, $x \in R^2$, be uniformly continuous and bounded from above and from below by positive constants. Let the Markov process $\xi^h(t)$ with values in R^2, if it is at a point x at time t, perform a jump of size h along the first coordinate with density $h^{-1}\lambda_1(t, x)$, of size $-h$ along the first coordinate with density $h^{-1}\lambda_{-1}(t, x)$, and of size h along the second coordinate with density $h^{-1}\lambda_2(t, x)$; between the jumps let it be constant. Then

$$G_0(t, x; z) = \lambda_1(t, x)(e^{z_1} - 1) + \lambda_{-1}(t, x)(e^{-z_1} - 1) + \lambda_2(t, x)(e^{z_2} - 1);$$

$$H_0(t, x; u) = u^1 \ln \frac{u^1 + \sqrt{(u^1)^2 + 4\lambda_1(t, x)\lambda_{-1}(t, x)}}{2\lambda_1(t, x)} +$$

$$+ \sqrt{(u^1)^2 + 4\lambda_1(t, x)\lambda_{-1}(t, x)} + \lambda_1(t, x) + \lambda_{-1}(t, x) +$$

$$+ u^2 \ln \frac{u^2}{\lambda_2(t, x)} - u^2 + \lambda_2(t, x).$$

Conditions **A** - **C**, **D'**, **D"**, **E'**, and **E"** are satisfied; in particular,

$$\{u: \underline{H}_0(u) < \infty\} = \{u = (u^1, u^2): u^2 \geq 0\},$$

and in Condition **D'** we can take $u_0 = (0, 1)$.

For the empirical distribution function (see Subsection 4.2.4) the conditions of Theorem 4.3.2 (and even the weaker conditions given in 4.2.3) are not satisfied, because the corresponding function $G_0(t, x; z) = \frac{1-x}{1-t}(e^z - 1)$ has a singularity at $t = 1$ and vanishes at $x = 1$. Still, let us perform the calculations:

$$H_0(t, x; u) = u \ln(1 - t) - u \ln(1 - x) + u \ln u - u + \frac{1-x}{1-t};$$

$$S_{0,1}(\phi) = \int_0^1 \dot{\phi}(t) \ln(1 - t)\, dt - \int_0^1 \dot{\phi}(t) \ln(1 - \phi(t))\, dt +$$

$$+ \int_0^1 \dot{\phi}(t) \ln \dot{\phi}(t)\, dt - \int_0^1 \dot{\phi}(t)\, dt + \int_0^1 \frac{1 - \phi(t)}{1 - t}\, dt = \int_0^1 \dot{\phi}(t) \ln \dot{\phi}(t)\, dt.$$

So, the functional $S_{0,1}(\phi)$ proves to be equal to the entropy of the distribution with distribution function $\phi(t)$, $0 \leq t \leq 1$, taken with the minus sign. The fact that $n \cdot S_{0,1}(\phi)$ is the action functional for the empirical distribution function as $n \to \infty$ does not follow from our results; it was established by I. N. Sanov [1].

Note that in the case of the empirical distribution function the generalized Cramér's transformation (see § 2.2) has a clear meaning. Calculations show that if we take the function $z^n(t, x)$ equal to

$$n\,[\ln(1-t) - \ln(1-\phi(t)) + \ln\dot{\phi}(t)]$$

(which does not depend on x), then the generalized Cramér's transformation consists in replacing the uniformly distributed random variables by variables with distribution function $\phi(t)$ (and the density $\pi(0, 1)$ proves to be equal to the likelihood ratio). One can base upon this fact proofs of theorems on large deviations for empirical distribution functions; various generalizations are possible (see, e.g., Bahadur [1]).

4.3.4. Diffusion processes with small diffusion

An important class of families of Markov processes depending on a parameter is that of diffusion processes with small diffusion in a Euclidean space or on a manifold.

Theorem 4.3.3. *Let a manifold of class $C^{(2)}$ with a distinguished atlas satisfying Condition λ be given. Let $(\xi^h(t), \mathrm{P}^h_{t,x})$, $0 \le t \le T$, for every $h > 0$, be the diffusion process on this manifold with generator A^h_t described in local coordinates by*

$$A^h_t f(x) = \sum_i b^{h,i}(t, x)\frac{\partial f}{\partial x^i} + \frac{h}{2}\sum_{i,j} a^{ij}(t, x)\frac{\partial^2 f}{\partial x^i \partial x^j},$$

where $b^{h,i}(t, x) \to b^i(t, x)$ uniformly with respect to t, with respect to x in the domain of a given chart, and with respect to all charts of the atlas chosen. Let $b^i(t, x)$, $a^{ij}(t, x)$ be bounded and continuous, uniformly with respect to all t, x and all charts of the atlas; let the inverse matrix $(a_{ij}(t, x)) = (a^{ij}(t, x))^{-1}$ have the same properties.

Then $h^{-1} S_{0,T}(\phi)$, where

$$S_{0,T}(\phi) = \int_b^T \frac{1}{2}\sum_{i,j} a_{ij}(t, \phi(t)) \times$$

$$\times (\dot{\phi}^i(t) - b^i(t, \phi(t)))\,(\dot{\phi}^j(t) - b^j(t, \phi(t)))\,dt,$$

is the action functional for the considered family of processes as $h \to 0$, uniformly with respect to the initial point.

The **proof** consists of using Theorem 3.3.2. ◊

The distinction between very large, not very large and super-large deviations, introduced in the previous section, is here inessential: this pattern falls into all three cases. The results on large deviations for diffusion processes with small diffusion were obtained by Freidlin, Wentzell [1], [4].

4.3.5. Theorem 4.3.4

Let X be a compact r-dimensional Riemannian manifold of class $C^{(2)}$. Let $(\xi^{h,h}(t), \mathsf{P}^{h,h}_{t,x})$, for each $h > 0$, be a Markov process of the form described in Subsection 4.2.6, with $\tau = h$; let $b^{h,h}(t, x) \to 0$ as $h \to 0$, uniformly with respect to t, x. Define the function $G_0(t, x; z)$ of $t \in [0, T)$, $x \in X$, $z \in T^*X_x$ by

$$G_0(t, x; z) = \ln \int_{TX_x} e^{zu} \mu_{t,x}(du),$$

and consider the function $H_0(t, x; u) \leftrightarrow G_0(t, x; z)$. For absolutely continuous $\phi(t)$, $0 \le t \le T$, with values in X we set

$$S_{0,T}(\phi) = \int_0^T H_0(t, \phi(t); \dot{\phi}(t))\, dt,$$

otherwise $S_{0,T}(\phi) = +\infty$.

Theorem 4.3.4. *Let the functions* G_0, H_0, *satisfy Conditions* **A** - **E**. *Then* $h^{-1} S_{0,T}(\phi)$ *is the action functional for the family of processes* $\xi^{h,h}(t)$ *as* $h \to 0$, *uniformly with respect to the initial point.*

Proof. Put $V = \{(x, y): \rho(x, y) < \rho_0\}$, where ρ_0 is a positive number. We have to verify Conditions (3.3.1'), (3.3.2) - (3.3.4) with h as the parameter θ, $\tau(\theta) = h$ and $k(\theta) = h^{-1}$. We will use Theorem 3.3.2'. Taking into account the way the process $\xi^{h,h}(t)$ is constructed, we have for t's that are multiples of h:

$$\mathsf{P}^{h,h}_{t,x}\{\rho(x, \xi^{h,h}(t+h)) \ge \rho_0\} = \mu_{t,x}(\overline{A^h_{t,x}}),$$

where

$$A^h_{t,x} = \{u \in TX_x : \rho(x, e(x, hb^{h,h}(t, x) + hu)) < \rho_0\}.$$

Since $\rho(x, e(x, u)) \le |u|_x$ (for small $|u_x|$ this inequality becomes an equality), we have

$$\overline{A^h_{t,x}} \subseteq \{u : |hb^{h,h}(t, x) + hu|_x \ge \rho_0\}.$$

For sufficiently small h we have $|hb^{h,h}(t, x)|_x < \rho_0/2$, and $\overline{A^h_{t,x}} \subseteq \{u: |u|_x \ge \rho_0/2h\}$. Hence

$$\mathsf{P}^{h,h}_{t,x}\{\rho(x, \xi^{h,h}(t+h)) \ge \rho_0\} \le \mu_{t,x}\{u: |u|_x \ge \rho_0/2h\}.$$

Now we choose a natural n, positive d, and, for each x, unit vectors $z(1), \ldots, z(n) \in T^*X_x$ so that the polyhedron $\{u: z(i)\, u < d,\ 1 \le i \le n\}$ lies inside the unit ball $\{u: |u|_x < 1\}$. Then for an arbitrary $a > 0$, by Chebyshev's inequality,

$$\mu_{t,x}\{u : |u|_x \geq \rho_0/2h\} \leq \sum_{i=1}^{n} \mu_{t,x}\{u : z(i)u \geq d\rho_0/2h\} \leq$$

$$\leq \sum_{i=1}^{n} \int_{TX_x} e^{az(i)u}\, \mu_{t,x}(du) \cdot e^{-ad\rho_0/2h} = \sum_{i=1}^{n} e^{G_0(t,x;\, az(i))}\, e^{-ad\rho_0/2h} \leq$$

$$\leq \sum_{i=1}^{n} e^{\overline{G}_0(az(i)A_x^{-1})}\, e^{-ad\rho_0/2h} \leq n \exp\{\max_{|z| \leq Ca} \overline{G}_0(z)\}\, e^{-ad\rho_0/2h}$$

Here C is the least upper bound of $|zA_x^{-1}|$ over all $x \in X$, $z \in T^*X_x$ with $|z|_x = 1$ and all charts of the chosen atlas on X; this least upper bound is finite by virtue of Condition λ.

It follows from these estimates that

$$\overline{\lim_{h \downarrow 0}}\, h \ln \sup_{t,x} \mathsf{P}_{t,x}^{h,h}\{\rho(x, \xi^{h,h}(t+h)) \geq \rho_0\} \leq -ad\rho_0/2h.$$

Since $a > 0$ is arbitrary, we have

$$\lim_{h \downarrow 0} h \ln \sup_{t,x} \mathsf{P}_{t,x}^{h,h}\{\rho(x, \xi^{h,h}(t+h)) \geq \rho_0\} = -\infty.$$

The additional factor h^{-1} under the logarithm sign does not influence the limit, and Condition (3.3.1') is satisfied.

Now we write down the truncated cumulant of the argument $h^{-1}z$ and its derivatives, multiplied by h:

$$hG_V^{h,h}(t, x; h^{-1}z) =$$

$$= \ln\left[\mu_{t,x}(\overline{A_{t,x}^h}) + \int_{A_{t,x}^h} \exp\{h^{-1}z\Delta\psi\}\, \mu_{t,x}(du)\right],$$

where

$$\Delta\psi = \psi(e(x, hb^{h,h}(t,x) + hu)) - \psi(x) = (\Delta\psi^1, \ldots, \Delta\psi^r).$$

Denote the expression between brackets by Ψ^h. Then we have:

$$\frac{\partial}{\partial z_i}(hG_V^{h,h}(t, x; h^{-1}z)) = (\Psi^h)^{-1} \int_{A_{t,x}^h} h^{-1}\Delta\psi^i e^{h^{-1}z\Delta\psi} d\mu_{t,x},$$

$$\frac{\partial^2}{\partial z_i \partial z_j}(hG_V^{h,h}(t, x; h^{-1}z)) = (\Psi^h)^{-1} \int_{A_{t,x}^h} h^{-2}\Delta\psi^i \Delta\psi^j e^{h^{-1}z\Delta\psi} d\mu_{t,x} +$$

$$+ (\Psi^h)^{-2} \int\limits_{A^h_{t,x}} h^{-1}\Delta\psi^i e^{h^{-1}z\Delta\psi} d\mu_{t,x} \int\limits_{A^h_{t,x}} h^{-1}\Delta\psi^j e^{h^{-1}z\Delta\psi} d\mu_{t,x} .$$

In these integrals $h^{-1}\Delta\psi \to A_x u$, $e^{h^{-1}z\Delta\psi} \to e^{zA_x u}$ as $h \downarrow 0$, uniformly with respect to $|z| \le z_0 < \infty$, $|u|_x \le u_0 < \infty$, all charts (W, ψ) of our atlas, and all x, $\rho(x, X\backslash W) > \lambda/4$. The uniform convergence of the integrals is ensured by the fact that $\sup_{t,x} \mu_{t,x}\{u: |u|_x \ge u_0\} \to 0$ as $u_0 \to \infty$ and that the integrands are dominated by the functions const $(1 + |u|_x^2)\exp\{\text{const}\cdot z_0 \cdot |u|_x\}$, which are uniformly integrable with respect to the measures $\mu_{t,x}$.

We find that the function $hG_V^{h,h}(t, x; h^{-1}z)$ and its derivatives converge uniformly, as $h \downarrow 0$, to $G_0(t, x; zA_x)$ and the derivatives of this function, and (3.3.2) - (3.3.4) are satisfied.

This proves the theorem. ◊

The result obtained is close to that of Azencott, Ruget [1], but Conditions **A** - **E** differ from their conditions. In particular, instead of the most complicated Condition **C** that, within the range of small values of H_0, amounts to the requirement of uniform continuity of this function, Azencott, Ruget [1] impose another restriction, which amounts to a Lipschitz condition for small values of H_0; on the other hand, within the range of large values of H_0 their conditions may be less restrictive.

We have seen that Conditions **A** - **E** are satisfied for the process of shifts along geodesics with uniform distribution in the direction and a distribution μ on the radius, satisfying the conditions of Subsection 4.1.2; the action functional has in this case the form

$$h^{-1}\int_0^T A_0(|\dot\phi(t)|_{\phi(t)})\,dt .$$

4.4. The case of not very large deviations

4.4.1. Theorem 4.4.1. *Let a family of Markov processes of one of the two classes described in Subsections* 4.2.1, 4.2.3 *be given. Let the function* G_* *(defined by formula* (4.2.2) *or* (4.2.6)) *be finite and bounded for all* t, x *and all sufficiently small* $|z|$; *let* $\nabla_z G_*(t, x; 0) \equiv 0$; *let the matrix*

$$(A^{ij}(t, x)) = \left(\frac{\partial^2 G_*}{\partial z_i \partial z_j}(t, x; 0)\right)$$

be bounded, uniformly positive definite and uniformly continuous with respect to t, x. Put

$$(A_{ij}(t, x)) = (A^{ij}(t, x))^{-1}. \tag{4.4.1}$$

Let $b^{\tau, h}(t, x) \to b(t, x)$ *uniformly with respect to t, x as* $\tau h^{-2} \to \infty$, $\tau h^{-1} \to 0$, *the function* $b(t, x)$ *being bounded and uniformly continuous with respect to t, x. Put*

$$H_0(t, x; u) = \frac{1}{2}\sum_{i,j} A_{ij}(t, x)(u^i - b^i(t, x))(u^j - b^j(t, x)). \tag{4.4.2}$$

Define the functional $S_{0,T}$ *by formula* (3.1.1).

Then $\tau h^{-2} S_{0,T}(\phi)$ *is the action functional for the family of processes* $(\xi^{\tau, h}(t), \mathrm{P}^{\tau, h}_{t, x})$ *as* $\tau h^{-2} \to \infty$, $\tau h^{-1} \to 0$, *uniformly with respect to the initial point.*

Proof. We use Theorem 3.2.3 or 3.2.2', taking for θ the vector parameter (τ, h), giving to $\theta \to$ the meaning $\tau h^{-2} \to \infty$, $\tau h^{-1} \to 0$, and putting $k(\theta) = \tau h^{-2}$ and, in the discrete case, $\tau(\theta) = \tau$.

It follows from the boundedness of G_* for small $|z|$ that, say, all third derivatives with respect to z are bounded for small $|z|$; and using the MacLaurin expansion for the function G_* we obtain that relations (3.2.1) - (3.2.3) are satisfied with

$$G_0(t, x; z) = zb(t, x) + \frac{1}{2}\sum_{i,j} A^{ij}(t, x) z_i z_j .$$

This function satisfies Conditions **A** - **E**, and its Legendre transform is H_0. This proves the theorem. ◊

4.4.2. The case of infinite exponential moments

We consider the case of discrete time. The condition $\nabla_z G_*(t, x; 0) \equiv 0$ can be rewritten as

$$\int u \, d\mu_{t, x} \equiv 0,$$

and the expression for $A^{ij}(t, x)$ as

$$A^{ij}(t, x) = \int_{R^r} u^i u^j \mu_{t,x}(du). \tag{4.4.3}$$

So we can define the functional $S_{0,T}(\phi)$ without supposing that exponential moments exist.

Let us show that $\tau h^{-2} S_{0,T}(\phi)$ remains the action functional for the considered family of processes without this supposition, but under the condition that τh^{-1} tends to zero sufficiently rapidly.

The subexponential case

Theorem 4.4.2. *Let $\mu_{t,x}\{u: |u| \geq y\} \leq \exp\{-Ky^{\beta}\}$ for all t, x, y, where $0 < \beta < 1$, $K > 0$; $\int_{R^r} u\,\mu_{t,x}(du) = 0$ for all t, x. Let the matrix $(A^{ij}(t, x))$ be bounded, uniformly positive definite and uniformly continuous with respect to t, x; let $b^{\tau,h}(t, x) \to b(t, x)$ uniformly with respect to t, x as $\tau h^{-2} \to \infty$, $\tau h^{-2+\beta} \to 0$, and let the function $b(t, x)$ be bounded and uniformly continuous with respect to t, x.*

Then $\tau h^{-2} S_{0,T}(\phi)$, where $S_{0,T}(\phi)$ is given by formulas (3.1.1), (4.4.3), (4.4.1), (4.4.2), *is the action functional for the family of processes $\xi^{h,h}(t)$ as $\tau h^{-2} \to \infty$, $\tau h^{-2+\beta} \to 0$, uniformly with respect to the initial point.*

Proof. We apply Theorem 3.3.2'. Put $V = \{(x, y): |x - y| < 1\}$; verify (3.3.1'):

$$\mathsf{P}^{\tau,h}_{t,x}\{(x, \xi^{\tau,h}(t + \tau)) \notin V\} = \mu_{t,x}\{u: |u - \tau h^{-1} b^{\tau,h}(t, x)| \geq h^{-1}\} \leq$$

$$\leq \mu_{t,x}\{u: |u| \geq h^{-1}/2\} \leq \exp\{-K(2h)^{-\beta}\}$$

for sufficiently large τh^{-2} and small $\tau h^{-2+\beta}$, from which

$$\tau^{-1} h^2 \ln\left[\tau^{-1} \sup_{t,x} \mathsf{P}^{\tau,h}_{t,x}\{(x, \xi^{\tau,h}(t + \tau)) \notin V\}\right] \leq$$

$$\leq \tau^{-1} h^2 (-\ln \tau - K(2h)^{-\beta}).$$

Since τ tends to zero slower than h^2, we have $|\ln \tau| \leq |\ln h^2|$, and this logarithm can be neglected as compared to the second term; the limit (3.3.1') is equal to $-\infty$.

Now we evaluate the truncated cumulant and verify (3.3.2) - (3.3.4). We have

$$(\tau h^{-2})^{-1} G_V^{\tau, h}(t, x; \tau h^{-2} z) =$$

$$= \tau^{-2} h^2 \ln \left[1 + \int_{|u - \tau h^{-1} b^{\tau, h}(t, x)| < h^{-1}} (\exp\{Q_{t, x}^{\tau, h}(z, u)\} - 1)\, \mu_{t, x}(du) \right],$$

where

$$Q_{t, x}^{\tau, h}(z, u) = \tau^2 h^{-2} z b^{\tau, h}(t, x) + \tau h^{-1} z u .$$

We have to prove that as $\tau h^{-2} \to \infty$, $\tau h^{-2+\beta} \to 0$, the above expression tends uniformly with respect to t, x, and z, $|z| \le z_0 = \text{const} < \infty$, to

$$G_0(t, x; z) = zb(t, x) + \frac{1}{2} \sum_{i, j} A^{ij}(t, x)\, z_i z_j .$$

It is sufficient to prove that the following expression converges uniformly to this limit:

$$\tau^{-2} h^2 \int_{|u - \tau h^{-1} b^{\tau, h}(t, x)| < h^{-1}} (\exp\{Q_{t, x}^{\tau, h}(z, u)\} - 1)\, \mu_{t, x}(du) =$$

$$= \int_{|u - \tau h^{-1} b^{\tau, h}(t, x)| < h^{-1}} \tau^{-2} h^2 (\exp\{Q_{t, x}^{\tau, h}(z, u)\} - 1 - Q_{t, x}^{\tau, h}(z, u))\, \mu_{t, x}(du) +$$

$$+ zb^{\tau, h}(t, x)\, \mu_{t, x}\{u: |u - \tau h^{-1} b^{\tau, h}(t, x)| < h^{-1}\} +$$

$$+ \tau^{-1} h \int_{|u - \tau h^{-1} b^{\tau, h}(t, x)| < h^{-1}} z u \mu_{t, x}(du).$$

The second term converges uniformly to $zb(t, x)$ by virtue of the uniform convergence $b^{\tau, h}(t, x) \to b(t, x)$ and the estimate

$$1 - \mu_{t, x}\{u: |u - \tau h^{-1} b^{\tau, h}(t, x)| < h^{-1}\} \le$$

$$\le 4h^2 \int |u|^2 \mu_{t, x}(du) = 4h^2 \sum_i A^{ii}(t, x) \le \text{const} \cdot h^2 \to 0.$$

The third term is equal to

$$- \tau^{-1} h \int_{|u - \tau h^{-1} b^{\tau, h}| \ge h^{-1}} z u \mu_{t, x}(du)$$

and does not exceed in absolute value

$$2\tau^{-1}h^2 z_0 \int |u|^2 \, d\mu_{t,x} \le \text{const} \cdot \tau^{-1}h^2 \to 0.$$

We split the first term into I_1, the integral over $|u| < \tau^{-1/4}h^{1/4}$, and I_2, the integral over the remaining u.

The function

$$\tau^{-2}h^2 (\exp \{Q_{t,x}^{\tau,h}(z,u)\} - 1 - Q_{t,x}^{\tau,h}(z,u))$$

converges to $(zu)^2/2$ uniformly in $|z| \le z_0$, $|u| < \tau^{-1/4}h^{1/4}$, and the integral I_1 converges uniformly to

$$\frac{1}{2}\sum_{i,j} A^{ij}(t,x)\, z_i z_j.$$

We estimate I_2 using the inequality $|e^a - 1 - a| \le e^{|a|} - 1 - |a| < e^{|a|}$: for $|z| \le z_0$,

$$|I_2| \le \tau^{-2}h^2 \exp \{\tau^2 h^{-2} z_0 |b^{\tau,h}(t,x)|\} \times$$

$$\times \int_{\tau^{-1/4}h^{1/4} \le |u| < 2h^{-1}} e^{\tau h^{-1} z_0 |u|} \mu_{t,x}(du).$$

The exponent outside the integral sign converges uniformly to 1. The last integral is equal to

$$\int_{[\tau^{-1/4}h^{1/4},\, 2h^{-1})} e^{\tau h^{-1} z_0 y} \, d(-\mu_{t,x}\{u : |u| \ge y\}) =$$

$$= -e^{\tau h^{-1} z_0 y} \mu_{t,x}\{u : |u| \ge y\} \Bigg|_{\tau^{-1/4}h^{1/4}}^{2h^{-1}} +$$

$$+ \int_{\tau^{-1}h^{1/4}}^{2h^{-1}} \mu_{t,x}\{u : |u| \ge y\} \, de^{\tau h^{-1} z_0 y} \le$$

$$\le e^{z_0 \tau^{3/4} h^{-3/4}} \exp \{-K(\tau^{-1/4}h^{1/4})^{\beta}\} +$$

$$+ \tau h^{-1} z_0 \int_{\tau^{-1/4} h^{1/4}}^{2h^{-1}} \exp\{-Ky^{\beta} + \tau h^{-1} z_0 y\}\, dy.$$

In the first term we have $\tau^{3/4} h^{-3/4} \to 0$, the first exponent tends to 1, and the entire term tends to zero at exponential rate. In the integral term obtained by integration by parts, the derivative with respect to y of the expression under the exponent sign is equal to

$$\tau h^{-1} z_0 - K\beta y^{\beta - 1} \le \tau h^{-1} z_0 - K\beta\,(h/2)^{1-\beta} < -Ch^{1-\beta},\ C > 0,$$

if only $\tau h^{-2+\beta}$ is sufficiently small. So this term does not exceed

$$\tau h^{-1} z_0 \exp\{-Ky^{\beta} + \tau h^{-1} z_0 y\}\ \Big|_{y = \tau^{-1/4} h^{1/4}} \times$$

$$\times \int_{\tau^{-1/4} h^{1/4}}^{2h^{-1}} \exp\{-Ch^{1-\beta}\,(y - \tau^{-1/4} h^{1/4})\}\, dy <$$

$$< \exp\{z_0 \tau^{3/4} h^{-3/4} - K\,(\tau^{-1/4} h^{1/4})^{\beta}\}.$$

Finally we obtain

$$|\, I_2 \,| \le \mathrm{const} \cdot \tau^{-2} h^2 \exp\{-K\,(\tau^{-1/4} h^{1/4})^{\beta}\} \to 0,$$

which proves (3.3.2).

In quite the same way we verify (3.3.3) and (3.3.4). The theorem is proved. ◊

The corresponding result for processes obtained starting from sums of independent random variables is proved in Mogul'skii [1], Theorem 1.

This result is analogous to results on large deviations of order $o\,(n^{\beta/(4-2\beta)})$ for normalized sums of n independent, identically distributed random variables X_i with

$$\mathsf{P}\,\{\,|X_i| \ge y\} \le e^{-Ky^{\beta}}$$

(cf. Ibragimov, Linnik [1], Theorem 13.1.1, where, of course, precise results and not rough ones as in our case are dealt with).

4.4.3. Theorem 4.4.2'

Results of the same kind can be obtained for families of processes with continuous time. Let $(\xi^{\tau, h}(t), \mathsf{P}^{\tau, h}_{t, x})$ be a family of processes of the form described in Subsection 4.2.3. Put

$$A^{ij}(t, x) = a^{ij}(t, x) + \int_{R^r} u^i u^j \nu_{t,x}(du). \qquad (4.4.4)$$

Theorem 4.4.2'. *Let* $\nu_{t,x}\{u: |u| \ge y\} \le \exp\{-Ky^{\beta}\}$ *for all* $t, x, y \ge y_0$, *where* $0 < \beta < 1$, $K > 0$. *Let the matrix* $(A^{ij}(t, x))$ *be bounded, uniformly positive-definite and uniformly continuous with respect to* t, x; *let* $b^{\tau,h}(t, x) \to b(t, x)$ *uniformly with respect to* t, x *as* $\tau h^{-2} \to \infty$, $\tau h^{-2+\beta} \to 0$, *and let the function* $b(t, x)$ *be bounded and uniformly continuous with respect to* t, x.

Then $\tau h^{-2} S_{0,T}(\phi)$, *where* $S_{0,T}$ *is given by formulas* (3.1.1), (4.4.4), (4.4.1), (4.4.2), *is the action functional for the family of processes* $(\xi^{\tau,h}(t), \mathsf{P}^{\tau,h}_{t,x})$ as $\tau h^{-2} \to \infty$, $\tau h^{-2+\beta} \to 0$, *uniformly with respect to the initial point.*

The **proof** is the same as that of Theorem 4.4.2, but somewhat simpler, because, in particular, one has to consider the set $\{u: |u| < h^{-1}\}$ instead of

$$\{u: |u - \tau h^{-1} b^{\tau,h}(t, x)| < h^{-1}\}. \ \lozenge$$

4.4.4. The case of power "tails"

Theorem 4.4.3. *Let* $(\xi^{\tau,h}(t), \mathsf{P}^{\tau,h}_{t,x})$ *be a family of Markov processes of one of the two classes described in Subsections* 4.2.1, 4.2.3; *let the function* $(A^{ij}(t, x))$ *be defined by* (4.4.3) *or* (4.4.4). *Let the conditions*

$$\mu_{t,x}\{u: |u| \ge y\} \le Cy^{-\beta}, \ \beta > 2, \ C > 0, \quad \int u \, d\mu_{t,x} \equiv 0$$

or the corresponding condition on $\nu_{t,x}$ *be satisfied. Let the matrix* $(A^{ij}(t, x))$ *be bounded, uniformly positive definite and uniformly continuous with respect to* t, x; *let* $b^{\tau,h}(t, x) \to b(t, x)$ *uniformly with respect to* t, x *as* $h \to 0$, $\tau h^{-2} \to \infty$, $\tau h^{-2} = o(\ln h^{-1})$, *and let the limiting function be bounded and uniformly continuous.*

Then $\tau h^{-2} S_{0,T}(\phi)$ *is the action functional for the family of processes considered as* $h \to 0$, $\tau h^{-2} \to \infty$, $\tau h^{-2} = o(\ln h^{-1})$, *uniformly with respect to the initial point.*

This time we will give the **proof** for the case of continuous time. We apply Theorem 3.3.2. Take

$$V = \{(x, y): |x - y| < 1\};$$

verify (3.3.1): for $h \to 0$, $\tau h^{-2} \to \infty$, $\tau h^{-2} = o(\ln h^{-1})$,

$$\tau^{-1}h^2 \sup_{t,\,x} \ln \lambda_{t,\,x}^{\tau,\,h} \{y\colon (x, y) \notin V\} \le \tau^{-1}h^2 \ln (Ch^{\beta}) \to -\infty.$$

Then we verify Condition (3.3.2). Find the truncated cumulant $G_V^{\tau,\,h}(t, x; z)$ and $\tau^{-1}h^2 G_V^{\tau,\,h}(t, x; \tau h^{-2}z)$:

$$\tau^{-1}h^2 G_V^{\tau,\,h}(t, x; \tau h^{-2}z) = zb_V^{\tau,\,h}(t, x) + \frac{1}{2}\sum_{i,\,j} a^{ij}(t, x)\, z_i z_j +$$

$$+ \tau^{-2}h^2 \int\limits_{|u| < h^{-1}} (e^{\tau h^{-1}zu} - 1 - \tau h^{-1}zu)\, \nu_{t,\,x}(du), \qquad (4.4.5)$$

$$b_V^{\tau,\,h}(t, x) = b^{\tau,\,h}(t, x) - \tau^{-1}h \int\limits_{|u| \ge h^{-1}} u\nu_{t,\,x}(du).$$

We have to prove that expression (4.4.5) converges, uniformly with respect to t, x and z, $|z| \le z_0 = \text{const} < \infty$, to

$$G_0(t, x; z) = zb(t, x) + \frac{1}{2}\sum_{i,\,j}\left[a^{ij}(t, x) + \int\limits_{R^r} u^i u^j \nu_{t,\,x}(du)\right] z_i z_j\,.$$

The difference between the linear terms contains the term with $b^{\tau,\,h}(t, x) - b(t, x)$, which converges uniformly to 0, and the scalar product of z by

$$\tau^{-1}h \int\limits_{|u| \ge h^{-1}} u\nu_{t,\,x}(du).$$

This integral does not exceed

$$\tau^{-1}h^2 \int |u|^2\, \nu_{t,\,x}(du) \le \tau^{-1}h^2 \sum_i A^{ii}(t, x) \le \text{const} \cdot \tau^{-1}h^2 \to 0.$$

The difference between the non-linear terms is equal to

$$\int\limits_{|u| < h^{-1}} \tau^{-2}h^2 \left[e^{\tau h^{-1}zu} - 1 - \tau h^{-1}zu - \frac{(\tau h^{-1}zu)^2}{2}\right] \nu_{t,\,x}(du) -$$

$$- \int\limits_{|u| > h^{-1}} \frac{(zu)^2}{2}\, \nu_{t,\,x}(du).$$

That the second integral tends uniformly to zero is clear at once; at first, just as in the proof of Theorem 4.4.2, we split into the integrals I_1 over $|u| < h^{-1}/\ln h^{-1}$ and I_2 over $h^{-1}/\ln h^{-1} \le |u| < h^{-1}$; and I_1 is estimated in the same way as in that proof. The integral I_2 is estimated as follows:

$$|I_2| \le \int\limits_{h^{-1}/\ln h^{-1} \le |u| < h^{-1}} \tau^{-2} h^2 e^{\tau h^{-1} z_0 |u|} \nu_{t,x}(du) \le$$

$$\le \tau^{-2} h^2 e^{z_0 \tau h^{-2}/\ln h^{-1}} \cdot C\,(h^{-1}/\ln h^{-1})^{-\beta} +$$

$$+ \tau^{-1} h z_0 \int\limits_{h^{-1}/\ln h^{-1}}^{h^{-1}} e^{\tau h^{-1} z_0 y}\, C y^{-\beta} dy.$$

Consider the first term: the exponent in it converges to 1, and the rest is equal to

$$C\,(\tau^{-1} h^2)^2 \cdot h^{\beta - 2} \ln^{\beta} h^{-1} \to 0.$$

The integrand in the second term is downward convex, so the integral is at most the length of the interval multiplied by half the sum of the values at its ends; that is, the second term doed not exceed

$$\tau^{-1} h z_0 \cdot h^{-1} e^{\tau h^{-2} z_0} \cdot C h^{\beta} \ln^{\beta} h^{-1} = C z_0\, \tau^{-1} h^2 h^{\beta - 2} e^{z_0 \tau h^{-2}} \ln^{\beta} h^{-1}.$$

For

$$h \to 0,\ \tau h^{-2} \to \infty,\ \tau h^{-2} = o\,(\ln h^{-1})$$

the exponent does not exceed $h^{-\gamma}$, where γ is an arbitrarily small positive number, and the whole expression tends to 0.

In verifying (3.3.3), (3.3.4) the following integrals arise:

$$\int\limits_{h^{-1}/\ln h^{-1} \le |u| < h^{-1}} \tau^{-1} h e^{\tau h^{-1} z_0 |u|} |u|\, \nu_{t,x}(du),$$

$$\int\limits_{h^{-1}/\ln h^{-1} \le |u| < h^{-1}} e^{\tau h^{-1} z_0 |u|} |u|^2\, \nu_{t,x}(du),$$

and the fact that they converge to zero is proved in a similar way. ◊

4.5. Some other patterns of not very large deviations

4.5.1. Theorem 4.5.1

Let $(\xi^{h,h}(t), \mathsf{P}^{h,h}_{t,x})$ be a family of Markov processes of one of the two classes described in Subsections 4.2.1 and 4.2.3 with $\tau = h$; we will suppose that $b^{h,h}(t, x) \equiv b(t, x)$, and in the discrete case in addition that

$$\int u \, d\mu_{t,x} \equiv 0.$$

Define the function G_* by formula (4.2. 2) or (4.2.6).

Suppose that the function $b(t, x)$ is continuous with respect to t, x and satisfies a Lipschitz condition in x. Denote by $X(t, x)$ the value at time t of the solution of the equation $\dot{x}(t) = b(t, x(t))$ with initial condition $x(0) = x$; $Y(t, x)$ will denote the function inverse to $X(t, x)$ in the second argument. The mapping Y "straightens" the paths of the process $\xi^{h,h}(t)$ that are most probable for small h.

Put, for $h > 0$, $\beta > 0$, $x_0 \in R^r$,

$$\eta^{h,\beta}_{x_0}(t) = \beta\,[Y(t, \xi^{h,h}(t)) - x_0].$$

We will obtain a theorem on the action functional for the family of processes $\eta^{h,\beta}_{x_0}(t)$ as $h \to 0$, $\beta \to \infty$, $h\beta^2 \to 0$.

Let $(A_{ij}(t, x))$ denote the inverse of the matrix of second-order derivatives with respect to z of the function G_* at $z = 0$. Put

$$C_{ij}(x_0; t) = \sum_{k,\,l} \frac{\partial X^k}{\partial x^i}(t, x_0)\, \frac{\partial X^l}{\partial x^j}(t, x_0)\, A_{kl}(t, X(t, x_0)); \qquad (4.5.1)$$

$$H^\eta_0(x_0; t; u) = \frac{1}{2} \sum_{i,\,j} C_{ij}(x_0; t)\, u^i u^j. \qquad (4.5.2)$$

For $x_0 \in R^r$ we define the functional $S^\eta_{0,T}(x_0; \phi)$ for absolutely continuous functions $\phi(t)$, $0 \le t \le T$, by

$$S^\eta_{0,T}(x_0; \phi) = \int_0^T H^\eta_0(x_0; t; \dot{\phi}(t))\, dt.$$

Theorem 4.5.1. *Let the function G_* be finite and bounded for all t, x and all sufficiently small z; let the matrix*

$$\left(\frac{\partial^2 G_*}{\partial z_i \partial z_j} (t, x; 0) \right)$$

be bounded, uniformly positive definite and uniformly continuous with respect to t, x; let the function b (t, x) be bounded and uniformly continuous together with its first- and second-order derivatives with respect to x.

Then the expression $h^{-1}\beta^{-2} S^{\eta}_{0,T}(x_0; \phi)$ *is the action functional for the family of processes* $\eta^{h,\beta}_{x_0}(t)$ *as* $\beta \to \infty$, $h\beta^2 \to 0$, *uniformly with respect to* $x_0 \in R^r$ *and the initial point of the process* $\eta^{h,\beta}_{x_0}$, *when changing over a bounded region; that is:*

a) *the functional* $S^{\eta}_{0,T}(x_0; \phi)$ *is lower semicontinuous in* ϕ, *uniformly with respect to* x_0;

b) *the functions of the set*

$$\bigcup_{x_0 \in R^r} \bigcup_{x \in R^r} \Phi^{\eta}_{x_0;\, x;\, [0, T]}(s)$$

are equicontinuous, where

$$\Phi^{\eta}_{x_0;\, x;\, [0, T]}(s) = \{\phi: \phi(0) = x,\ S^{\eta}_{0,T}(x_0; \phi) \le s\};$$

c) *for any positive* δ, γ, s_0, *for any bounded set* $K \subset R^r$, *for sufficiently large* β *and sufficiently small* $h\beta^2$, *for arbitrary* $x_0 \in R^r$, $x \in K$ *and* $\phi \in \Phi^{\eta}_{x_0;\, x;\, [0, T]}(s_0)$,

$$\mathsf{P}^{h,\beta}_{x_0;\, 0,\, x}\{\rho_{0,T}(\eta^{h,\eta}_{x_0}, \phi) < \delta\} \ge \exp\{-h^{-1}\beta^{-2}[S^{\eta}_{0,T}(x_0; \phi) + \gamma]\},$$

where $\mathsf{P}^{h,\beta}_{x_0;\, 0,\, x}$ *denotes the probability evaluated under the assumption that* $\eta^{h,\beta}_{x_0}(0) = x$ *(i.e.,* $\xi^{h,h}(0) = x_0 + \beta^{-1}x$*);*

d) *for any* $\delta > 0$, $\gamma > 0$, $s_0 > 0$, *bounded* $K \subset R^r$, *for sufficiently large* β *and small* $h\beta^2$, *for any* $x_0 \in R^r$, $x \in K$, $s \le s_0$,

$$\mathsf{P}^{h,\beta}_{x_0;\, 0,\, x}\{\rho_{0,T}(\eta^{h,\beta}_{x_0}, \Phi^{\eta}_{x_0;\, x;\, [0, T]}(s)) \ge \delta\} \le \exp\{-h^{-1}\beta^{-2}(s - \gamma)\}.$$

We will carry out the **proof** for the continuous case. We will use Theorem 3.2.3

and the remarks to Theorems 3.2.3, 3.2.3'. To this end we first of all find local characteristics of the process $\eta_{x_0}^{h,\beta}(t)$.

The compensating operator $\mathfrak{A}^h$ of the process $\xi^{h,h}(t)$ is given by the formula

$$\mathfrak{A}^h f(t,x) = \frac{\partial f}{\partial t}(t,x) + \sum_i b^i(t,x)\frac{\partial f}{\partial x^i}(t,x) + \frac{h}{2}\sum_{i,j} a^{ij}(t,x)\frac{\partial^2 f}{\partial x^i \partial x^j}(t,x) +$$

$$+ h^{-1}\int\left[f(t, x+hu) - f(t,x) - h\sum_i \frac{\partial f}{\partial x^i}(t,x)\,u^i\right]\nu_{t,x}(du).$$

Using this expression we can find the compensator of the random function

$$f(t, \eta_{x_0}^{h,\beta}(t)) = F(t, \xi^{h,h}(t)),$$

where

$$F(t,x) = f(t, \beta\,[Y(t,x) - x_0]).$$

Calculation shows that the compensating operator $\mathfrak{A}_{x_0}^{h,\beta}$ of the Markov process $\eta_{x_0}^{h,\beta}(t)$ is given by

$$\mathfrak{A}_{x_0}^{h,\beta} f(t,x) = \frac{\partial f}{\partial t}(t,x) + \beta\sum_i \frac{\partial f}{\partial x^i}(t,x)\left\{\frac{\partial Y^i}{\partial t}(t, X(t, x_0 + \beta^{-1}x)) +\right.$$

$$+ \sum_j b^j(t, X(t, x_0 + \beta^{-1}x))\,\frac{\partial Y^i}{\partial x^j}(t, X(t, x_0 + \beta^{-1}x)) +$$

$$+ \frac{h}{2}\sum_{k,l} a^{kl}(t, X(t, x_0 + \beta^{-1}x))\,\frac{\partial^2 Y^i}{\partial x^k \partial x^l}(t, X(t, x_0 + \beta^{-1}x)) +$$

$$+ h^{-1}\int\Big[Y^i(t, X(t, x_0 + \beta^{-1}x) + hu) - Y^i(t, X(t, x_0 + \beta^{-1}x)) -$$

$$\left.- h\sum_j \frac{\partial Y^i}{\partial x^j}(t, X(t, x_0 + \beta^{-1}x))\,u^j\Big]\,\nu_{t, X(t, x_0 + \beta^{-1}x)}(du)\right\} +$$

$$+ \frac{h\beta^2}{2}\sum_{i,j}\frac{\partial^2 f}{\partial x^i \partial x^j}(t,x)\cdot\sum_{k,l}\frac{\partial Y^i}{\partial x^k}(t, X(t, x_0 + \beta^{-1}x)) \times$$

$$\times \frac{\partial Y^j}{\partial x^l}(t, X(t, x_0 + \beta^{-1}x))\,a^{kl}(t, X(t, x_0 + \beta^{-1}x)) +$$

$$+ h^{-1}\int\Big[f(t, \beta\,[Y(t, X(t, x_0 + \beta^{-1}x) + hu) - x_0]) - f(t,x) -$$

$$-\sum_i \frac{\partial f}{\partial x^i}(t,x)\,(\beta\ \ [Y^i(t, X(t, x_0+\beta^{-1}x)+hu) - x_0{}^i] - x^i)\Big]\ \nu_{t,\, X(t,\, x_0+\beta^{-1}x)}(du).$$

The first two terms between brackets cancel one another by virtue of the definition of the functions X, Y. Denote the sum over k, l in the third term between brackets by III^i, the integral in the fourth by IV^i, the sum over k, l in the first term after the brackets by V^{ij}. Put

$$\mathrm{VI}^i = Y^i(t, X(t, x_0+\beta^{-1}x)+hu) - (x_0^i + \beta^{-1}x^i).$$

We write out the expression for $h\beta^2 G_{x_0}^{h,\beta}(t,\ x;\ h^{-1}\beta^{-2}z)$, where $G_{x_0}^{h,\beta}$ is the cumulant of the process $\eta_{x_0}^{h,\beta}$:

$$h\beta^2 G_{x_0}^{h,\beta}(t, x; h^{-1}\beta^{-2}z) = \sum_i z_i\,\{h\beta\,\mathrm{III}^i + h^{-1}\beta\,\mathrm{IV}^i\} +$$

$$+\frac{1}{2}\sum_{i,j} z_i z_j\,\mathrm{V}^{ij} + \int \beta^2\left(e^{h^{-1}\beta^{-1}\sum_i z_i \mathrm{VI}^i} - 1 - h^{-1}\beta^{-1}\sum_i z_i\,\mathrm{VI}^i\right)\times$$

$$\times\ \nu_{t,\, X(t,\, x_0+\beta^{-1}x)}(du). \qquad (4.5.3)$$

The first order derivatives with respect to z_i of this function at zero are equal to $h\beta\mathrm{III}^i + h^{-1}\beta\mathrm{VI}^i$; we write out the second order derivatives:

$$\frac{\partial^2}{\partial z_i \partial z_j}(h\beta^2 G_{x_0}^{h,\beta}(t,\ x;\ h^{-1}\beta^{-2}z)) =$$

$$= \mathrm{V}^{ij} + \int h^{-2} e^{h^{-1}\beta^{-1}\sum_i z_i\,\mathrm{VI}^i}\cdot \mathrm{VI}^i\mathrm{VI}^j\,\nu_{t,\, X(t,\, x_0+\beta^{-1}x)}(du). \quad (4.5.4)$$

Put

$$C^{ij}(x_0; t) = \sum_{k,l}\frac{\partial Y^i}{\partial x^k}(t, X(t, x_0))\,\frac{\partial Y^j}{\partial x^l}(t, X(t, x_0))\,\frac{\partial^2 G_*}{\partial z_k \partial z_l}(t, X(t, x_0); 0),$$

$$G_0^{\eta}(x_0;\ t,\ z) = \frac{1}{2}\sum_{i,j} C^{ij}(x_0;\ t)\, z_i z_j\,.$$

We will prove that the first order derivatives of the expression (4.5.3) at zero and its second order derivatives at other points converge, uniformly in every bounded region, to the corresponding derivatives of the function G_0^{η}.

The integrand in IV^i can be expressed by means of Taylor expansion in the form

$$\frac{h^2}{2} \sum_{j,k} \frac{\partial^2 Y^i}{\partial x^j \partial x^k}(t, \tilde{x}) \cdot u^j u^k;$$

from the boundedness of $G_*(t, x; z)$ for small z it follows that the integral is $O(h^2)$, uniformly with respect to t, x_0 and x. Since the expression III^i is clearly bounded, $h\beta\, III^i + h^{-1}\beta\, IV^i = O(h\beta)$, and the first order derivatives of

$$h\beta^2 G_{x_0}^{h,\beta}(t, x; h^{-1}\beta^2 z)$$

at $z = 0$ converge to 0 as $\beta \to \infty$, $h\beta^2 \to 0$ (as it was to be proved).

To find the limit of the expression (4.5.4), we again use Taylor expansion

$$VI^i = h \sum_k \frac{\partial Y^i}{\partial x^k}(t, X(t, x_0 + \beta^{-1}x) + h\tilde{u})\, u^k.$$

We see that the integrand in (4.5.4) converges to

$$\sum_{k,l} \frac{\partial Y^i}{\partial x^k}(t, X(t, x_0)) \frac{\partial Y^j}{\partial x^l}(t, X(t, x_0))\, u^k u^l$$

as $\beta \to \infty$, $h\beta^2 \to 0$. For $\beta \geq \beta_0$, $|z| \leq C$ it is dominated by the function

$$\text{const} \cdot |u|^2 \exp\left\{\beta_0^{-1} C r^2 \sup_{i,k,t,x} \left|\frac{\partial Y^i}{\partial x^k}\right| |u|\right\}.$$

For sufficiently large β_0 this function is uniformly integrable with respect to all measures $\nu_{t,x}$, and limit transition under the integral sign is possible. Hence we obtain that for sufficiently large β and small $h\beta^2$, uniformly in all t, x_0 and in x and z varying in bounded regions, the expression (4.5.4) is close to

$$\sum_{k,l} \frac{\partial Y^i}{\partial x^k}(t, X(t, x_0)) \frac{\partial Y^j}{\partial x^l}(t, X(t, x_0)) \times$$

$$\times \left[a^{kl}(t, X(t, x_0 + \beta^{-1}x)) + \int u^k u^l \nu_{t, X(t, x_0 + \beta^{-1}x)}(du)\right].$$

The expression between brackets is equal to

$$\frac{\partial^2 G_*}{\partial z_k \partial z_l}(t, X(t, x_0 + \beta^{-1}x); 0),$$

and as $\beta \to \infty$ it converges to

$$\frac{\partial^2 G_*}{\partial z_k \partial z_l}(t, X(t, x_0); 0)$$

(by the assumed uniform continuity of $\frac{\partial^2 G_*}{\partial z_k \partial z_l}$ with respect to the second argument).

So convergence of the expression (4.5.3) and its derivatives to the function G_0^{η} and its derivatives has been proved; the Legendre transform of G_0^{η} is the function H_0^{η} given by the formulas (4.5.1), (4.5.2); applying Theorem 3.2.3 and the remarks to it gives the statement required. ◊

4.5.2. Theorem 4.5.2

Theorem 4.5.1 means that for the family of processes $\eta_{x_0}^{h,\beta}(t)$ the action functional is the same as for the family of Gaussian processes with independent increments with zero drift. Another way of "straightening" most probable paths, under which only one path is straightened, consists in subtracting the solution $X(t, x_0)$ of the equation $\dot{x}(t) = b(t, x(t))$ from the process.

For the family of stochastic processes

$$\zeta_{x_0}^{h,\beta}(t) = \beta\,[\xi^{h,h}(t) - X(t, x_0)]$$

the action functional is the same as for the family of Gaussian Markov processes with linear drift.

Put

$$B_k^i(x_0; t) = \frac{\partial b^i}{\partial x^k}(t, X(t, x_0));$$

$$H_0^{\zeta}(x_0; t, x; u) = \frac{1}{2} \sum_{i,j} A_{ij}(t, X(t, x_0)) \times$$

$$\times \left(u^i - \sum_k B_k^i(x_0; t)\, x^k\right)\left(u^j - \sum_l B_l^j(x_0; t)\, x^l\right);$$

$$S_{0,T}^{\zeta}(x_0; \phi) = \int_0^T H_0^{\zeta}(x_0; t, \phi(t); \dot{\phi}(t))\, dt.$$

Theorem 4.5.2. *Under the conditions of Theorem* 4.5.1 *the expression*

$$h^{-1}\beta^{-2}\, S_{0,T}^{\zeta}(x_0; \phi)$$

is the action functional for the family of processes $\zeta_{x_0}^{h,\beta}(t)$ *as* $\beta \to \infty$, $h\beta^2 \to 0$, *uniformly with respect to* $x_0 \in R^r$ *and the initial point, as it varies in every bounded region.*

The **proof** can be carried out in the same way as that of the previous theorem; or one can deduce the assertion of the theorem from Theorem 4.5.1. Indeed, Taylor expansion gives us

$$\zeta_{x_0}^{h,\beta;i}(t) = \sum_k \frac{\partial X^i}{\partial x^k}(t, x_0) \cdot \eta_{x_0}^{h,\beta;k}(t) + O(\beta^{-1} | \eta_{x_0}^{h,\beta}(t) |^2).$$

Thus, up to infinitesimals, one random function is obtained starting from the other by means of a non-degenerate linear transformation depending on t; and the necessary recalculation of the action functional is carried out in an elementary way. ◊

4.6. The case of super-large deviations

4.6.1. Formal calculations for the case of super-large deviations

As already mentioned (Subsection 4.2.1), results on super-large deviations for sums of independent random variables are highly sensitive to the nature of decrease of the "tails" of their distributions. In this connection it is of some interest to consider a process $\xi^{\tau,h}$ whose jumps have no non-trivial "tails", i.e. are bounded in absolute value by a constant multiplied by h (recall that the sizes of jumps in the pattern of families of Markov stochastic processes described in § 4.2 are proportional to h). It goes without saying that non-zero large deviation probabilities can only arise here in the continuous case, because of an exceedingly large number of small jumps.

We will consider the one-dimensional case only.

Let $(\xi^{\tau,h}(t), \mathsf{P}_{t,x}^{\tau,h})$ be a family of processes of the form described in Subsection 4.2.3, with cumulant given by

$$G^{\tau,h}(t, x; z) = zb^{\tau,h}(t, x) + \tau^{-1}G_*(t, x; hz),$$

$$G_*(t, x; z) = \int (e^{zu} - 1 - zu)\, u^{-2} \mu_{t,x}(du).$$

Suppose that the measure $\mu_{t,x}$ is bounded in t, x and, for every t, x, is concentrated on the interval $[u_-(t, x), u_+(t, x)]$ where the functions $u_-(t, x)$, $u_+(t, x)$ are uniformly continuous, and $u_-(t, x) < -c$, $u_+(t, x) > c$, $c > 0$. We introduce a condition meaning that the ends of the interval $[u_-(t, x), u_+(t, x)]$ are, uniformly with respect to t, x, "sufficiently filled" with the measure $\mu_{t,x}$ (cf. Subsection 4.1.3): for any $\kappa > 0$ there exists a $\lambda > 0$ such that for all t, x,

$$\mu_{t,x}\,[u_-(t,x), (1-\kappa)\, u_-(t,x)] > \lambda,$$

$$\mu_{t,x}\,[(1-\kappa)\, u_+(t,x),\; u_+(t,x)] > \lambda.$$

We require the functions $b^{\tau,h}(t, x)$ to converge, uniformly with respect to t, x, to a uniformly continuous and bounded function $b(t, x)$ as $h \to 0$, $\tau^{-1}h \to 0$.

Calculations show that the Legendre transform $H^{\tau,h}(t, x; u)$ of the cumulant $G^{\tau,h}(t, x; z)$ has order $h^{-1}\ln(\tau h^{-1})$ as $h \to 0$, $\tau^{-1}h \to 0$. Therefore we have to take $k(\tau, h) = h^{-1}\ln(\tau h^{-1})$.

The limit, as $h \to 0$, $\tau^{-1}h \to 0$, of the expression

$$k(\tau, h)^{-1}\, G^{\tau,h}(t, x; k(\tau, h)\, z) = zb^{\tau,h}(t, x) + \tau^{-1}h\ln^{-1}(\tau h^{-1}) \times$$
$$\times \int \left(e^{\ln(\tau h^{-1})\, zu} - 1 - \ln(\tau h^{-1})\, zu\right) u^{-2}\mu_{t,x}(du)$$

as can be easily verified, is equal to $zb(t, x)$ for $u_-(t, x)^{-1} < z < u_+(t, x)^{-1}$, and to $+\infty$ for $z > u_+(t, x)^{-1}$ and $z < u_-(t, x)^{-1}$. The conditions of Theorem 3.2.3 are not fulfilled. Nevertheless, we put

$$G_0(t, x; z) = \begin{cases} zb(t, x) & u_-(t, x)^{-1} \le z \le u_+(t, x)^{-1}; \\ +\infty & \text{for remaining } z, \end{cases}$$

and find the Legendre transform of this function:

$$H_0(t, x; u) = \begin{cases} \dfrac{u - b(t, x)}{u_+(t, x)} & \text{for } u \ge b(t, x); \\[2ex] \dfrac{u - b(t, x)}{u_-(t, x)} & \text{for } u \le b(t, x). \end{cases}$$

We define the functional

$$S_{0,T}(\phi) = \int_0^T H_0(t, \phi(t);\; \dot{\phi}(t))\, dt$$

(as usual, $S_{0,T}(\phi)$ is assumed to be $+\infty$ for not absolutely continuous ϕ).

4.6.2. Comparison with Poisson variables

The form of the functions $k(\tau, h)$, $H_0(t, x; u)$ and of the functional $S_{0,T}(\phi)$ is to be compared with those obtained in the simple case of summation of independent Poisson variables. Namely, the following assertion holds:

Let λ and c be positive constants, let $\xi^{\tau, h}$ be a Poisson variable with parameter $\tau^{-1}\lambda$, multiplied by ch (or a sum of independent random variables having the prescribed value of the parameter); then for $0 \le a < b \le \infty$,

$$\lim_{\substack{h \to 0 \\ \tau^{-1} h \to 0}} (h^{-1} \ln (\tau h^{-1}))^{-1} \ln \mathsf{P} \{\xi^{\tau, h} \in [a, b]\} = - \min_{x \in [a, b]} S(x), \qquad (4.6.1)$$

where $S(x) = x/c$ (so that the limit (4.6.1) *is equal to a/c).*

The **proof** is elementary.

The rôle played here by the Poisson distribution is clear; the jumps of a locally infinitely divisible process with jump sizes within a certain set can be regarded as occuring according to a "Poisson flow of events", made inhomogeneous in time and space.

4.6.3. The functional $S_{0,T}(\phi)$ does not satisfy the conditions necessary

However, the functional $S_{0,T}(\phi)$ cannot claim to be the normalized action functional for our family of processes, because the set

$$\Phi_{x;[0,T]}(s) = \{\phi : \phi(0) = x, S_{0,T}(\phi) \le s\}$$

is not compact: there do exist non-absolutely continuous functions ϕ such that

$$\varliminf_{\psi \to \phi} S_{0,T}(\psi) < \infty.$$

Even if we redefine the functional for such functions, replacing its value by the lower limit, the set $\Phi_{x;[0,T]}(s)$ does not become compact. Perhaps this can be repaired by considering, instead of the metric

$$\rho_{0,T}(\phi, \psi) = \sup_{0 \le t \le T} |\phi(t) - \psi(t)|,$$

another, "shorter", metric, e.g., the Skorohod metric, as it is done in Mogul'skii [1]; but our basic estimates (§ 2.2, § 2.3) are adjusted to the metric $\rho_{0,T}$ and they yield nothing for a new metric.

Nevertheless, we can obtain some partial results.

4.6.4. Theorem 4.6.1. *Under the above-mentioned conditions, for any positive δ and γ, for any equicontinuous set Φ of functions, for sufficiently small h and $\tau^{-1}h$, for every $x_0 \in R^1$, and any function $\phi \in \Phi$, $\phi(0) = x_0$,*

$$\mathsf{P}^{\tau, h}_{0, x_0} \{\rho_{0,T}(\xi^{\tau, h}, \phi) < \delta\} \ge \exp\{-h^{-1} \ln(\tau h^{-1}) [S_{0,T}(\phi) + \gamma]\}.$$

The **proof** is based on Theorem 2.2.1. ◊

Note that the condition $h \to 0$, $\tau^{-1}h \to 0$ does not impose any restrictions on the variation of τ: this parameter can, for small h and $\tau^{-1}h$, take arbitrarily small as well as arbitrarily large values, and all intermediate values.

4.6.5. Theorem 4.6.2

As to the converse estimate, the secondary terms in (2.3.13) prove to be small as compared to the principal term, not in the whole range of variation of τ and h described by the requirements $h \to 0$, $\tau^{-1}h \to 0$, but only in a narrow part of this range. Indeed, in order that the first term at the right-hand side of (2.3.13) should be negligible, we have to choose the set U_0 ever increasing; to ensure that

$$\delta' \geq \Delta t_{\max} \sup \{ |u| : u \in U_0 \}$$

one must choose a partition $0 = t_0 < t_1 < \dots < t_n = T$ with $\Delta t_{\max}$ tending to 0; and this increases the term N^n. A result can only be obtained if $h \to 0$ and $\tau^{-1}h$ tends to 0, but slower than any positive power of h. Since it follows from $\tau^{-1}h \to 0$, $h^{\alpha} = o(\tau^{-1}h)$, for small $\alpha > 0$, that $h \to 0$, $\tau \to 0$, the additional requirement $h \to 0$ is superfluous.

Theorem 4.6.2. *Under the same assumptions we have, for any positive δ, γ and s_0, positive ε and α such that for all positive τ, h with $\tau^{-1}h \leq \varepsilon$, $\tau^{-1}h \geq h^{\alpha}$, for any $x_0 \in R^1$, and any $s \leq s_0$,*

$$\mathsf{P}^{\tau, h}_{0, x_0} \{\rho_{0,T} (\xi^{\tau, h}, \Phi_{x_0; [0, T]}(s)) \geq \delta\} \leq \exp\{-h^{-1} \ln(\tau h^{-1})(s - \gamma)\}.$$

The **proof** is obtained applying Theorem 2.3.2. ◊

CHAPTER 5

PRECISE ASYMPTOTICS FOR LARGE DEVIATIONS

5.1. The case of the Wiener process

5.1.1. Formulation of Theorem 5.1.1

Let $\xi^h(t)$ be a family of stochastic processes depending on a positive parameter h; let the probabilities of low-probability events related to this process be described, as $h \downarrow 0$, by the action functional $h^{-1} S(\phi)$. If F is a measurable continuous bounded functional, the mean $\mathsf{M}^h \exp\{h^{-1} F(\xi^h)\}$ is logarithmically equivalent to

$$\exp\{h^{-1} \max[F(\phi) - S(\phi)]\}$$

(see Introduction). One cannot expect to find precise asymptotics for this expectation without introducing additional restrictions. We will introduce such restrictions in this chapter. First, we will consider only a certain form of dependence of the Markov process $\xi^h(t)$ on the parameter; secondly, the functional F will be supposed to be smooth; thirdly, we will impose restrictions on the way in which the functional $F - S$ approaches its maximum. Namely, under our conditions the functional S will be twice differentiable on a certain space. At the maximum point the second derivative of $F - S$ is a negative semi-definite functional; we will suppose it to be strictly negative definite. The results obtained will be valid for asymptotics of expectations of the form

$$\mathsf{M}^h G(\xi^h) \exp\{h^{-1} F(\xi^h)\}$$

as well.

In this section we consider, after Schilder [1], the particular case when ξ^h is an r-dimensional Wiener process with variance of each coordinate equal to h per time unit. We can obtain such a process by taking $\xi^h(t) = h^{1/2} w(t)$ where $w(t)$ is the standard r-dimensional Wiener process. We will consider the process starting at point 0 at zero time; the probability $\mathsf{P}^h_{0,0}$ and expectation $\mathsf{M}^h_{0,0}$ will be denoted simply by P, M.

We know that the action functional for the family $\xi^h(t)$, $0 \le t \le T$, as $h \downarrow 0$, is given by $h^{-1} S(\phi)$, where $S(\phi)$ is the following quadratic functional:

$$S(\phi) = S_{0,T}(\phi) = \int_0^T \frac{1}{2} |\dot{\phi}(t)|^2 \, dt$$

for absolutely continuous ϕ, and $S(\phi) = +\infty$ otherwise.

Along with the space $\boldsymbol{C}\,[0, T]$ of continuous functions on $[0, T]$ with values in R^r we will consider the space $\boldsymbol{W}^{1,2}\,[0, T]$ of absolutely continuous functions with square integrable derivative, and the space $\boldsymbol{C}^1\,[0, T]$ of continuously differentiable functions; $\boldsymbol{C}_0\,[0, T]$, $\boldsymbol{W}_0^{1,2}\,[0, T]$ and $\boldsymbol{C}_0^1\,[0, T]$ will denote the subspaces of these spaces consisting of the functions equal to 0 for $t = 0$. We will use the notation

$$\|\phi\| = \max_{0 \le t \le T} |\phi(t)|.$$

Theorem 5.1.1. *Let* $\xi^h(t) = h^{1/2} w(t)$, *where* $w(t)$ *is the standard r-dimensional Wiener process starting at* $t = 0$ *at the point 0. Let F be a bounded continuous functional on the space* $\boldsymbol{C}_0\,[0, T]$. *Let the maximum* $\max\{F(\phi) - S(\phi): \phi \in \boldsymbol{C}_0\,[0, T]\}$ *be attained at a unique function* $\phi_0 \in \boldsymbol{C}_0\,[0, T]$. *Let the functional F be twice Fréchet differentiable at the point* ϕ_0; *let the second derivative of F at this point - the bilinear functional* $F''(\phi_0)(x_1, x_2)$ *- be such that for any non-zero function* $x \in \boldsymbol{C}_0\,[0, T]$ *the strict inequality*

$$\frac{1}{2} F''(\phi_0)(x, x) < S(x) \tag{5.1.1}$$

holds.

Then, as $h \downarrow 0$,

$$\mathsf{M} \exp\{h^{-1} F(\xi^h)\} \sim K_0 \exp\{h^{-1}[F(\phi_0) - S(\phi)]\}, \tag{5.1.2.}$$

where

$$K_0 = \mathsf{M} \exp\{\frac{1}{2} F''(\phi_0)(w, w)\}. \tag{5.1.3}$$

5.1.2. Proof. The linear term

By virtue of Lemma 0.1 (see Introduction), for an arbitrary $\varepsilon > 0$ there exists a $\gamma > 0$ such that for $h \downarrow 0$,

$$\mathsf{M} \exp\{h^{-1} F(\xi^h)\} = \mathsf{M}\{\rho_{0,T}(\xi^h, \phi_0) < \varepsilon;\ \exp\{h^{-1} F(\xi^h)\}\} +$$
$$+ o(\exp\{h^{-1}[F(\phi_0) - S(\phi_0) - \gamma]\}). \tag{5.1.4}$$

We use the generalized Cramér's transformation corresponding to the function $z^h(t) = h^{-1}(z_1(t), \ldots, z_r(t))$, which does not depend on x, where $z_i(t) = \dot{\phi}_0^i(t)$. This transformation takes the process $\xi^h(t)$ to a Wiener process with variance of each coordinate equal to h per time unit and with drift $\dot{\phi}_0(t)$; in other words, with respect to the new probability measure

$$\mathsf{P}^{z^h}(A) = \mathsf{M}\,(A; \exp\{\int_0^T z^h(t)\, d\xi^h(t) - h^{-1}\int_0^T \frac{1}{2}|\phi_0(t)|^2\, dt\})$$

the stochastic process $\xi^h(t)$ has the same distribution in the space $C_0\,[0, T]$ as $\phi_0(t) + h^{1/2} w(t)$.

We rewrite the expectation at the right side of (5.1.4) using the generalized Cramér's transformation:

$$\mathsf{M}\,\{\rho_{0,T}(\xi^h, \phi_0) < \varepsilon; \exp\{h^{-1} F(\xi^h)\}\} =$$

$$= \mathsf{M}^{z^h}\{\rho_{0,T}(\xi^h, \phi_0) < \varepsilon; \exp\{h^{-1} F(\xi^h) -$$

$$- h^{-1}\int_0^T \sum_{i=1}^r \dot{\phi}_0^i(t)\, d\xi^{hi}(t) + h^{-1}\int_0^T \frac{1}{2}|\dot{\phi}_0(t)|^2\, dt\}\} =$$

$$= \mathsf{M}\,\{\,\|w\| < h^{-1}\varepsilon; \exp\{h^{-1} F(\phi_0 + h^{1/2} w) -$$

$$- h^{-1/2}\int_0^T \sum_{i=1}^r \dot{\phi}_0^i(t)\, dw^i(t) - h^{-1}\int_0^T \frac{1}{2}|\dot{\phi}_0(t)|^2\, dt\}\}. \tag{5.1.5}$$

The last integral here has the minus sign, because a term

$$-h^{-1}\int_0^T |\dot{\phi}_0(t)|^2\, dt$$

arises from the stochastic integral.

It turns out that if the functional F is differentiable, the stochastic integral in this formula coincides with probability one with $F'(\phi_0)(w)$. Indeed, the first derivative of F at the point ϕ_0 - a linear functional on $C_0\,[0, T]$ - can be represented in the form

$$F'(\phi_0)(x) = \int_{(0,T]} \sum_{i=1}^r x^i(t)\, dV_i(t), \tag{5.1.6}$$

where $V_i(t)$ are right-continuous functions of bounded variation on $[0, T]$; we standardize them by the condition $V_i(T) = 0$. The quadratic functional S is also differentiable - of course not on the whole space $C_0\,[0, T]$, but only on the space $W_0^{1,2}\,[0, T]$ on which it takes finite values; for $x \in W_0^{1,2}\,[0, T]$,

$$S'(\phi_0)(x) = \int_0^T \sum_{i=1}^r \dot{\phi}_0^i(t)\, \dot{x}^i(t)\, dt.$$

Since the functional $F - S$ has an extremum at the point ϕ_0, its derivatives must vanish for all $x \in C_0^1[0, T]$:

$$\int_{(0,T]} \sum_{i=1}^r x^i(t)\, dV_i(t) - \int_0^T \sum_{i=1}^r \dot{\phi}_0^i(t)\, \dot{x}^i(t)\, dt = 0. \tag{5.1.7}$$

Integrate by parts the first integral (the side term arising vanishes):

$$-\int_0^T \sum_{i=1}^r V_i(t)\, \dot{x}^i(t)\, dt - \int_0^T \sum_{i=1}^r \dot{\phi}_0^i(t)\, \dot{x}^i(t)\, dt = 0. \tag{5.1.8}$$

We can take $\dot{x}^i(t)$ to be arbitrary continuous functions. Hence, it follows from (5.1.8) that $V_i(t) = -\dot{\phi}_0^i(t)$ for almost all t.

In the integral $F'(\phi_0)(w)$, too, we perform integration by parts, obtaining a stochastic integral equal to it almost surely:

$$F'(\phi_0)(w) = \int_{(0,T]} \sum_{i=1}^r w^i(t)\, dV_i(t) =$$

$$= -\int_0^T \sum_{i=1}^r V_i(t)\, dw^i(t) = \int_0^T \sum_{i=1}^r \dot{\phi}_0^i(t)\, dw^i(t) \tag{5.1.9}$$

(the side term arising vanishes since $w^i(0) = 0$, $V_i(T) = 0$).

Taking this into account, we rewrite formulas (5.1.4), (5.1.5) in the following form (the factors $\exp\{h^{-1} F(\phi_0)\}$ and $\exp\{-h^{-1} S(\phi_0)\}$ are taken out of the expectation sign):

$$\mathsf{M} \exp\{h^{-1} F(\xi^h)\} = \exp\{h^{-1}[F(\phi_0) - S(\phi_0)]\} \times$$
$$\times \mathsf{M}\chi_{\{\|w\| < h^{-1/2}\varepsilon\}} \exp\{h^{-1}[F(\phi_0 + h^{1/2} w) - F(\phi_0) - h^{1/2} F'(\phi_0)(w)]\} +$$
$$+ o(\exp\{h^{-1}[F(\phi_0) - S(\phi_0) - \gamma]\}). \tag{5.1.10}$$

5.1.3. Uniform integrability. End of the proof of Theorem 5.1.1

We write the Taylor expansion of the functional F near the function ϕ_0 up to second-order terms:

$$F(\phi_0 + h^{1/2}w) = F(\phi_0) + h^{1/2} F'(\phi_0)(w) + \frac{h}{2} F''(\phi_0)(w, w) + R(h^{1/2}w),$$

where $R(x) = o(\|x\|^2)$ as $x \to 0$. Substituting this into (5.1.10) we obtain

$$\mathsf{M} \exp\{h^{-1} F(\xi^h)\} = \exp\{h^{-1}[F(\phi_0) - S(\phi_0)]\} \times$$

$$\times \left[\mathsf{M}\chi_{\{\|w\| < h^{-1/2}\varepsilon\}} \exp\left(\frac{1}{2} F''(\phi_0)(w, w) + h^{-1} R(h^{1/2}w)\right) + o(e^{-\gamma h^{-1}}) \right]. \quad (5.1.11)$$

The random variable under the expectation sign converges to

$$\exp\{\frac{1}{2} F''(\phi_0)(w, w)\}$$

as $h \downarrow 0$. In order to establish the finiteness of the expectation of this limit (i.e., of the expectation (5.1.3)) and the equivalence (5.1.2), it is sufficient to establish for some $\varepsilon > 0$ that the random variables under the expectation sign in (5.1.11) are uniformly integrable for all $h > 0$.

Lemma 5.1.1. *Let $Q(x)$ be a continuous quadratic functional on $\mathbf{C}_0[0, T]$, and let $Q(x) < S(x)$ for all non-zero $x \in \mathbf{C}_0[0, T]$. Then there exists a $\kappa > 0$ such that*

$$Q(x) \le (1 - \kappa) S(x) \quad (5.1.12)$$

for all $x \in \mathbf{C}_0[0, T]$.

Proof. The set

$$\Phi_0(1) = \{x \in \mathbf{C}_0[0, T]: S(x) \le 1\}$$

is compact, so the continuous function $Q(x)$ attains its maximum on $\Phi_0(1)$ at some point x_0. If $x_0 = 0$, we can take 1 for κ, and if $x_0 \ne 0$, then $Q(x_0) < S(x_0) \le 1$, and we take $\kappa = 1 - Q(x_0)$. For an arbitrary non-zero $x \in \mathbf{C}_0[0, T]$ either $S(x) = +\infty$ and (5.1.12) is satisfied trivially; or $S(x) < \infty$, in which case we define $\tilde{x} = x / \sqrt{S(x)} \in \Phi_0(1)$ and have:

$$Q(x) = Q(\sqrt{S(x)} \cdot \tilde{x}) = S(x) \cdot Q(\tilde{x}) \le S(x)(1 - \kappa). \ \Diamond$$

Using this lemma we choose a $\kappa > 0$ for the functional $Q(x) = \frac{1}{2} F''(\phi_0)(x, x)$ which is less than $S(x)$ for $x \ne 0$ according to (5.1.1). Then we choose $\varepsilon > 0$ so that $R(x) \le \frac{\kappa}{4T} \|x\|^2$ for $\|x\| < \varepsilon$. The function under the expectation sign at the right-hand side of (5.1.11) does not exceed $\exp\{Q_1(w)\}$, where

$$Q_1(x) = \frac{1}{2} F''(\phi_0)(x, x) + \frac{\kappa}{4T} \|x\|^2 \le (1 - \kappa/2) S(x) \quad (5.1.13)$$

(since $\| x \|^2 \le 2TS(x)$). To establish the uniform integrability it is sufficient to verify that $\mathsf{M} \exp \{Q_1(w)\} < \infty$.

To this end we choose $\alpha > 0$ (arbitrary as yet) and consider for all natural m the $\alpha\sqrt{m}$-neighbourhoods of the sets

$$\Phi_0(m) = \{x \in C_0[0, T]: S(x) \le m\}.$$

We denote these neighbourhoods by $\Phi_0(m)_{+\alpha\sqrt{m}}$. The whole space $C_0[0, T]$ decomposes into the union

$$\Phi_0(1)_{+\alpha} \cup (\Phi_0(2)_{+\alpha\sqrt{2}} \setminus \Phi_0(1)_{+\alpha}) \cup \ldots$$

$$\ldots \cup (\Phi_0(m+1)_{+\alpha\sqrt{m+1}} \setminus \Phi_0(m)_{+\alpha\sqrt{m}}) \cup \ldots .$$

We estimate the parts of the integral of $\exp\{Q_1(w)\}$ corresponding to each of these sets:

$$\mathsf{M}\{w \in \Phi_0(m+1)_{+\alpha\sqrt{m+1}} \setminus \Phi_0(m)_{+\alpha\sqrt{m}};\ \exp\{Q_1(w)\}\} \le$$

$$\le \mathsf{P}\{w \notin \Phi_0(m)_{+\alpha\sqrt{m}}\} \exp\{\sup\{Q_1(x): x \in \Phi_0(m+1)_{+\alpha\sqrt{m+1}}\}\}. \quad (5.1.14)$$

For any function $x \in \Phi_0(m+1)$ there exists a function $\tilde{x} \in \Phi_0(m+1)$ such that $\| x - \tilde{x} \| < \alpha\sqrt{m+1}$. Since Q is a continuous quadratic functional, there exists a constant C such that

$$Q(x) \le Q(\tilde{x}) + C \| \tilde{x} + x \| \cdot \| x - \tilde{x} \| \le (1-\kappa)(m+1) +$$
$$+ C(\| \tilde{x} \| \cdot \alpha\sqrt{m+1} + 2(\alpha\sqrt{m+1})^2) \le$$
$$\le (m+1)(1 - \kappa + C\sqrt{2T}\alpha + 2C\alpha^2).$$

The term $\frac{\kappa}{4T}\| x \|^2$ in (5.1.13) does not exceed

$$\frac{\kappa}{4T}(\| \tilde{x} \| + \alpha\sqrt{m+1})^2 \le \frac{\kappa}{4T}\left(\sqrt{2TS(\tilde{x})} + \alpha\sqrt{m+1}\right)^2 \le$$

$$\le (m+1) \cdot \frac{\kappa}{4T}(\sqrt{2T} + \alpha)^2 = (m+1)\,\kappa\left(\frac{1}{2} + \frac{\alpha}{\sqrt{2T}} + \frac{\alpha^2}{4T}\right).$$

Thus, for $x \in \Phi_0(m+1)_{+\alpha\sqrt{m+1}}$ we have

$$Q_1(x) \le (m+1)\left(1 - \frac{\kappa}{2} + \frac{\kappa\alpha}{\sqrt{2T}} + \frac{\kappa\alpha^2}{T} + C\sqrt{2T}\,\alpha + 2C\alpha^2\right).$$

Choose $\alpha > 0$ in such a way that the right-hand side does not exceed $(m+1)(1-\kappa/3)$. Then

$$\sup\left\{Q_1(x): x \in \Phi_0(m+1)_{+\alpha\sqrt{m+1}}\right\} \le (m+1)(1-\kappa/3).$$

The probability at the right side of (5.1.14) is estimated using Theorem 2.3.1. Here the functional S plays the rôle not only of the normalized action functional for the family of processes ξ^h, but also of the action functional of the stochastic process w. We have

$$G(t, x; z) \equiv G(z) = \frac{1}{2} |z|^2, \quad H(t, x; u) \equiv H(u) = \frac{1}{2} |u|^2.$$

For δ' we take $\alpha\sqrt{m}/3$; we put

$$t_i = iT/n, \quad \Delta t_{\min} = \Delta t_{\max} = T/n;$$

$$k = 2r, \quad z(j) = \sqrt{m}\, z_0(j), \quad d(j) = \sqrt{m}\, d_0,$$

where $z_0(1) = (Z, 0, \ldots, 0)$, $z_0(2) = (-Z, 0, \ldots, 0)$, ..., $z_0(2r) = (0, \ldots, 0, -Z)$, $d_0 = Z^2$; and U_0 is the cube with side $2Z\sqrt{m}$ and centre at 0. We approximate the downward convex function $H(u) = \frac{1}{2} |u|^2$ on the cube

$$\{u: |u^1| < 1, \ldots, |u^r| < 1\}$$

by the polyhedron circumscribed from below,

$$\max_{1 \le j \le N} \left[z_0\{j\}\, u - \frac{1}{2} |z_0\{j\}|^2 \right],$$

with accuracy $\kappa/6T$. Take $z(j) = \sqrt{m}\, z_0\{j\}$. For ε_1, ε_2 we can take, respectively, 0, $m\kappa/6T$. In order that the inequality

$$\delta' = \alpha\sqrt{m}/3 \ge \frac{T}{n} \cdot \sup\{ |u| : u \in U_0\} = \frac{T}{n} Z\sqrt{r}\sqrt{m}$$

be satisfied, we have to take

$$n \ge 3Z\sqrt{r}\, T/\alpha. \tag{5.1.15}$$

Then the estimate (2.3.3) gives:

$$\mathsf{P}\{w \notin \Phi_0(m)_{+\alpha\sqrt{m}}\} = \mathsf{P}\{\rho_{0,T}(w, \Phi_0(m)) \ge \alpha\sqrt{m}\} \le$$

$$\le 2n \cdot 2r \cdot \exp\left\{\frac{T}{n}\left[\frac{1}{2} mZ^2 - mZ^2\right]\right\} + N^n \cdot \exp\{-m + T \cdot m\kappa/6T\} =$$

$$= 4nr \exp\left\{-m \frac{TZ^2}{2n}\right\} + N^n \exp\{-m(1 - \kappa/6)\}.$$

If $Z^2T \ge 2n$, we can neglect the first term at the right-hand side. This inequality and inequality (5.1.15) can be satisfied if we take a natural number $n \ge 18rT/\alpha^2$ and $Z = \sqrt{2n/T}$.

For this choice of the elements of our construction we obtain:

$$\mathsf{M} \exp \{Q_1(w)\} \le \sum_{m=0}^{\infty} \exp \{(m+1)(1-\kappa/3)\} \times$$

$$\times [4nr \exp \{-m\} + N^n \exp \{-m(1-\kappa/6)\}] \le$$

$$\le \exp \{1-\kappa/3\} [4nr + N^n] \frac{1}{1-e^{-\kappa/6}}.$$

This completes the proof of the theorem. ◊

5.1.4. Several identical maxima, or a factor $G(\xi^h)$. Unbounded functionals

In a hardly more complicated way we can examine the case when the absolute maximum of $F(\phi) - S(\phi)$ is attained not at one but at a finite number of points, and also the case when $\exp\{h^{-1}F(\xi^h)\}$ is multiplied by $G(\xi^h)$, where G is a continuous functional.

Theorem 5.1.2. *Let F be a bounded continuous functional; let the maximum $\max\{F(\phi) - S(\phi): \phi \in C_0[0, T]\}$ be attained at functions $\phi_1, \ldots, \phi_k \in C_0[0, T]$, and let the functional F be twice differentiable at these points with*

$$\frac{1}{2} F''(\phi_j)(x, x) < S(x)$$

for all non-zero $x \in C_0[0, T]$, $j = 1, \ldots, k$. Let G be a bounded measurable functional on $C_0[0, T]$ that is continuous at the points ϕ_j. Then, as $h \downarrow 0$,

$$\mathsf{M}G(\xi^h) \exp\{h^{-1}F(\xi^h)\} = \exp\{h^{-1} \max[F(\phi) - S(\phi)]\} \times$$

$$\times \left[\sum_{j=1}^{k} G(\phi_j)\, \mathsf{M} \exp\left\{\frac{1}{2} F''(\phi_j)(w, w)\right\} + o(1)\right]. \qquad (5.1.16)$$

The result obtained can be generalized to certain unbounded functionals F and G; it is sufficient to impose some growth conditions, for example: $F(\phi) = o(\|\phi\|^2)$ as $\|\phi\| \to \infty$, $G(\phi) = O(e^{C\|\phi\|^2})$ for any $C > 0$.

5.1.5. Refinements

Now let the functional F be differentiable s times, and G, $s-2$ times. Taking the Taylor expansion of F up to the s-th term, and of G up to the $(s-2)$-th, we can obtain more precise expressions than (5.1.16).

For simplicity of notation, we return to the case of a unique maximum.

The formula corresponding to (5.1.10) is as follows:

$$\mathsf{M}G(\xi^h)\exp\{h^{-1}F(\xi^h)\} = \exp\{h^{-1}[F(\varphi_0) - S(\varphi_0)]\}\times$$
$$\times[\mathsf{M}\chi_{\{\|w\|<h^{-1/2}\varepsilon\}}\,G(\varphi_0 + h^{1/2}w)\times$$
$$\times\exp\{h^{-1}[F(\varphi_0 + h^{1/2}w) - F(\varphi_0) - h^{1/2}F'(\varphi_0)(w)]\} +$$
$$+\,o(e^{-h^{-1}\gamma})]. \qquad (5.1.17)$$

Split the expression involving the factor $\chi_{\{\|w\|<h^{-1/2}\varepsilon\}}$ into two parts: the part with $\|w\| < h^{-0.1}$ and that with $h^{-0.1} \le \|w\| < h^{-1/2}\varepsilon$. The second part is estimated from above by the least upper bound of $|G|$ multiplied by

$$\mathsf{M}\left\{\|w\| \le h^{-0.1};\ \exp\left\{\frac{1}{2}F''(\varphi_0)(w, w) + \frac{\kappa}{4T}\|w\|^2\right\}\right\} \le$$
$$\le \sum_{m=m_0}^{\infty}\mathsf{M}\Big\{w \in \Phi_0(m+1)_{+\alpha\sqrt{m+1}} \setminus \Phi_0(m)_{+\alpha\sqrt{m}};$$
$$\exp\left\{\frac{1}{2}F''(\varphi_0)(w, w) + \frac{\kappa}{4T}\|w\|^2\right\}\Big\} \le$$
$$\le \exp\{1 - \kappa/3\}\,[4nr + N^n]\,\frac{e^{-m_0\kappa/6}}{1 - e^{-\kappa/6}}. \qquad (5.1.18)$$

Here m_0 is the integral part of $h^{-0.2}/(\sqrt{2T} + \alpha)^2$; the values of m less than m_0 are not compatible with the inequality $\|w\| \ge h^{-0.1}$, because

$$\sup\{\|x\|: x \in \Phi_0(m)_{+\alpha\sqrt{m}}\} \le \sqrt{m}\cdot(\sqrt{2T} + \alpha).$$

The expression (5.1.18) converges to 0 faster than any power of h as $h \downarrow 0$.

For the part of the expectation with $\|w\| < h^{-0.1}$, we use the fact that

$$F(\varphi_0 + x) = F(\varphi_0) + F'(\varphi_0)(x) + \frac{1}{2}F''(\varphi_0)(x, x) + O(\|x\|^3)$$

as $\|x\| \to 0$; so for $\|w\| < h^{-0.1}$ we have

$$h^{-1}\left[F(\varphi_0 + h^{1/2}w) - F(\varphi_0) - h^{1/2}F'(\varphi_0)(w) - \frac{h}{2}F''(\varphi_0)(w, w)\right] =$$
$$= O(h^{1/2}\|w\|^3) = O(h^{0.2}) \to 0.$$

To the left-hand side of this formula we apply the expansion

$$e^a = 1 + a + \frac{a^2}{2} + \ldots + \frac{a^{s-2}}{(s-2)!} + o(a^{s-2});$$

at the same time we take the Taylor expansion of the functional F up to terms of at least order s:

$$\exp\{h^{-1}[F(\phi_0+h^{1/2}w)-F(\phi_0)-h^{1/2}F'(\phi_0)\ (w)]\} =$$

$$= \exp\left\{\frac{1}{2}F''(\phi_0)\ (w, w)\right\}\cdot\left[1+\frac{h^{1/2}}{6}F'''(\phi_0)\ (w, w, w) + \right.$$

$$+\frac{h}{24}F^{\mathrm{IV}}(\phi_0)\ (w, w, w, w) + \frac{h}{72}(F'''(\phi_0)\ (w, w, w))^2 +$$

$$\left. + \ldots + o\left(h^{\frac{s-2}{2}}\left(\|w\|^s + \|w\|^{3(s-2)}\right)\right)\right].$$

Combining this with the expansion for G:

$$G(\phi_0+h^{1/2}w) = G(\phi_0) + h^{1/2}G'(\phi_0)\ (w) +$$

$$+\frac{h}{2}G''(\phi_0)\ (w, w) + \ldots + \frac{h^{\frac{(s-2)}{2}}}{(s-2)!}G^{(s-2)}(\phi_0)\ (w, \ldots, w) + o\left(h^{\frac{(s-2)}{2}}\|w\|^{s-2}\right),$$

we obtain

$$G(\phi_0+h^{1/2}w)\exp\{h^{-1}[F(\phi_0+h^{1/2}w)-F(\phi_0)-h^{1/2}F'(\phi_0)\ (w)]\} =$$

$$= \exp\left(\frac{1}{2}F''(\phi_0)\ (w, w)\right)\times$$

$$\times\left[G(\phi_0)+h^{1/2}G'(\phi_0)\ (w)+\frac{h^{1/2}}{6}G(\phi_0)\,F'''(\phi_0)\ (w, w, w) + \right.$$

$$+\frac{h}{2}G''(\phi_0)\ (w, w)+\frac{h}{6}G'(\phi_0)\ (w)\ F'''(\phi_0)\ (w, w, w) +$$

$$+\frac{h}{24}G(\phi_0)\,F^{\mathrm{IV}}(\phi_0)\ (w, w, w, w)+\frac{h}{72}G(\phi_0)\ (F'''(\phi_0)\ (w, w, w))^2 + \ldots$$

$$\left.\ldots + o\left(h^{\frac{s-2}{2}}\left(\|w\|^{s-2} + \|w\|^{3(s-2)}\right)\right)\right].$$

This expression is integrated over the set $\{\ \|w\| < h^{-0.1}\}$, which fills the whole space as h decreases. The expectations

$$\mathsf{M}\exp\left\{\frac{1}{2}F''(\phi_0)\ (w, w)\right\}\ G^{(k)}(\phi_0)\ (w, \ldots, w)\prod_i F^{(k_i)}(\phi_0)\ (w, \ldots, w)$$

are finite because the factor at the exponent does not exceed

$$\mathrm{const}\ \|w\|^l \le \mathrm{const}'\ \exp\left\{0.1\,\frac{\kappa}{T}\|w\|^2\right\}.$$

The integrals over $\{\ \|w\| < h^{-0.1}\}$ converge to their limits at exponential rate. Finally, the term

$$o\left(h^{\frac{s-2}{2}}\left(\|w\|^{s-2}+\|w\|^{3(s-2)}\right)\right)$$

yields, after multiplication by $\exp\{\frac{1}{2}F''(\phi_0)(w,w)\}$ and integration, $o\left(h^{\frac{s-2}{2}}\right)$.

So

$$MG(\xi^h)\exp\{h^{-1}F(\xi^h)\}=\exp\{h^{-1}[F(\phi_0)-S(\phi_0)]\}\times$$
$$\times\Big[G(\phi_0)+h^{1/2}\,\mathrm{M}\,(G'(\phi_0)(w)+\frac{1}{6}G(\phi_0)F'''(\phi_0)(w,w,w))+$$
$$+h\mathrm{M}\Big(\frac{1}{2}G''(\phi_0)(w,w)+\frac{1}{6}G'(\phi_0)(w)F'''(\phi_0)(w,w,w)+$$
$$+\frac{1}{24}G(\phi_0)F^{\mathrm{IV}}(\phi_0)(w,w,w,w)+$$
$$+\frac{1}{72}G(\phi_0)(F'''(\phi_0)(w,w,w))^2\Big)+\ldots+o\left(h^{\frac{s-2}{2}}\right)\Big]. \tag{5.1.19}$$

The expectations of the products of the exponent by the multilinear functionals of w of odd order vanish because the distribution of w is symmetric; i.e., in (5.1.19) only the terms with integral powers of h remain.

Theorem 5.1.3. *Let the conditions of Theorem* 5.1.1 *be satisfied. Let, in addition, the functional G be bounded, continuous and differentiable s* – 2 *times at the point* ϕ_0, *and let F be s times differentiable at* ϕ_0. *Then, as* $h \downarrow 0$,

$$MG(\xi^h)\exp\{h^{-1}F(\xi^h)\}=$$
$$=\exp\{h^{-1}[F(\phi_0)-S(\phi_0)]\}\left[\sum_{0\le i\le\frac{s-2}{2}}K_ih^i+o\left(h^{\frac{s-2}{2}}\right)\right], \tag{5.1.20}$$

where the coefficients K_i *are determined by the derivatives of G up to order* $2i$ *and those of F up to order* $2i+2$ *at* ϕ_0; *in particular,*

$$K_0=G(\phi_0)\,\mathrm{M}\exp\left\{\frac{1}{2}F''(\phi_0)(w,w)\right\},$$

$$K_1=\mathrm{M}\exp\left\{\frac{1}{2}F''(\phi_0)(w,w)\right\}\Big[\frac{1}{2}G''(\phi_0)(w,w)+$$
$$+\frac{1}{6}G'(\phi_0)(w)F'''(\phi_0)(w,w,w)+\frac{1}{24}G(\phi_0)F^{\mathrm{IV}}(\phi_0)(w,w,w,w)+$$
$$+\frac{1}{72}G(\phi_0)(F'''(\phi_0)(w,w,w))^2\Big].$$

5.2. Processes with frequent small jumps

In this section we will, after Dubrovskii [1], [2], study the case when $\xi^h(t) = \xi^{h,h}(t)$ is a family of processes of the class considered in § 4.3. The formulation and proof of the principal result will be similar to those of Theorem 5.1.1, but some new difficulties arise.

5.2.1. Some lemmas on derivatives

To begin with we set forth some auxiliary results concerning differentiation.

Lemma 5.2.1. *Let the function $H_0(t, x; u)$, $t \in [0, T]$, $x, u \in R^r$, be twice differentiable with respect to the pair x, u, and let the function itself and its derivatives be continuous with respect to t, x, u. Then the functional*

$$S(\varphi) = S_{0,T}(\varphi) = \int_0^T H_0(t, \varphi(t); \dot{\varphi}(t))\, dt \tag{5.2.1}$$

is twice Fréchet differentiable at every point $\varphi \in C^1[0, T]$ in the space $C^1[0, T]$. The first and second derivatives are given by the formulas

$$S'(\varphi)(x) = \int_0^T \sum_i \left[\frac{\partial H_0}{\partial x^i} x^i(t) + \frac{\partial H_0}{\partial u^i} \dot{x}^i(t)\right] dt \tag{5.2.2}$$

and

$$S''(\varphi)(x, x) = \int_0^T \sum_{i,j} \left[\frac{\partial^2 H_0}{\partial x^i \partial x^j} x^i(t)\, x^j(t) + \right.$$

$$\left. + 2\,\frac{\partial^2 H_0}{\partial x^i \partial u^j} x^i(t)\, \dot{x}^j(t) + \frac{\partial^2 H_0}{\partial u^i \partial u^j} \dot{x}^i(t)\, \dot{x}^j(t)\right] dt, \tag{5.2.3}$$

where the partial derivatives are taken at the point $(t, \varphi(t); \dot{\varphi}(t))$.

Lemma 5.2.2. *Let the functions $H_0(t, x; u)$, $G_0(t, x; z)$ be coupled by the Legendre transformation in the third argument and let them be twice continuously differentiable with respect to the two last arguments. Then the partial derivatives of the function G_0 at the point $(t, x; \nabla_u H_0(t, x; u))$ and of H_0 at the point $(t, x; u)$ are connected by the relations*

$$\frac{\partial G_0}{\partial x^i} = -\frac{\partial H_0}{\partial x^i},$$

$$\frac{\partial^2 H_0}{\partial x^i \partial u^j} = -\sum_k \frac{\partial^2 G_0}{\partial x^i \partial z_k} \frac{\partial^2 H_0}{\partial u^k \partial u^{i'}},$$

$$\frac{\partial^2 H_0}{\partial x^i \partial x^j} = -\frac{\partial^2 G_0}{\partial x^i \partial x^i} + \sum_{k,\,l} \frac{\partial^2 H_0}{\partial u^k \partial u^l} \frac{\partial^2 G_0}{\partial z_k \partial x^i} \frac{\partial^2 G_0}{\partial z_i \partial x^{j'}};$$

the matrix $\left(\frac{\partial^2 H_0}{\partial u^i \partial u^j}\right)$ *is the inverse of* $\left(\frac{\partial^2 G_0}{\partial z_i \partial z_j}\right)$.

The **proof** of both lemmas is straightforward differentation of the functional, or, respectively, of the formula

$$H_0(t, x; u) = \sum_k u^k \frac{\partial H_0}{\partial u^k}(t, x; u) - G_0(t, x; \nabla_u H_0(t, x; u)).\Diamond$$

Let $\boldsymbol{D}\,[0, T]$ be the space of functions on $[0, T]$ with values in R^r, continuous on the right and with limits on the left, with norm

$$\| x \| = \sup_{0 \le t \le T} | x(t) |;$$

and let $\boldsymbol{D}_0\,[0, T]$ be its subspace consisting of functions vanishing at $t = 0$.

The first derivative of a functional on $\boldsymbol{D}_0\,[0, T]$ is a continuous linear functional on this space.

Lemma 5.2.3. *Any continuous linear functional* $l(x)$ *on* $\boldsymbol{D}\,[0, T]$ *is representable in the form*

$$l(x) = \int_{(0,\,T]} \sum_{i=1}^{r} x^i(t)\, dV_i(t) + \sum_{j=1}^{\infty} \sum_{i=1}^{r} \alpha_i^j \cdot (x^i(t_j) - x^i(t_j -)),$$

where $V_i(t)$, $i = 1, \ldots, r$, *are right-continuous functions of bounded variation on* $[0, T]$, $t_j \in (0, T]$, $\sum_{j,\,i} |\alpha_i^j| < \infty$.

Proof. Consider the set $\mathfrak{T}$ consisting of points of the form $t \in [0, T]$ and $t-$, where $t \in (0, T]$; we introduce on $\mathfrak{T}$ the natural order and topology, taking as neighbourhoods all intervals (a, b), $a, b \in \mathfrak{T}$. In this topology $\mathfrak{T}$ is a compactum. The space $\boldsymbol{D}\,[0, T]$ with norm $\| \cdot \|$ can be identified with the space $\boldsymbol{C}\,(\mathfrak{T})$ of continuous functions on $\mathfrak{T}$. According to the Riesz theorem (see Halmos [1], § 56), continuous linear functionals on $\boldsymbol{C}\,(\mathfrak{T})$ can be represented as integrals with respect to charges (signed measures) on the σ-algebra of Borel subsets of $\mathfrak{T}$ (consisting of Borel sets of pairs t, $t-$ plus at most countable sets of individual points t_k, $t_k' -$). Every such charge is uniquely determined by its distribution function $V(\tau)$, $t \in \mathfrak{T}$, having at most a countable set of discontinuity points, and the series of sizes of the jumps at these points converges absolutely. Hence we obtain the assertion of the lemma. ◊

5.2.2. Localization, the generalized Cramér's transformation. Taylor expansion

Theorem 5.2.1. *Let* $(\xi^{h,h}(t), \mathsf{P}^{h,h}_{t,x})$, $t \in [0, T]$, $h > 0$, *be a family of locally infinitely divisible processes with cumulants* $G^{h,h}(t, x; z) = h^{-1} G_0(t, x; hz)$. *Let the functions* $G_0(t, x; z) \leftrightarrow H_0(t, x; u)$ *be everywhere finite and satisfy Conditions* **A** - **E** *of* § 3.1, 3.2; *let, in addition, the following conditions be satisfied:*

F. *The functions* $G_0(t, x; z)$ *and* $H_0(t, x; u)$ *are twice differentiable with respect to* $(x; z)$ *or, respectively, with respect to* $(x; u)$, *and their first and second derivatives are continuous with respect to* $(t, x; z)$, $(t, x; u)$.

G. *There exists a positive number b such that*

$$\sum_{i,j} \frac{\partial^2 G_0}{\partial z_i \partial z_j}(t, x; z)\, z_i z_j \geq b\, |z|^2$$

for all t, x, z.

Let the functional $S(\phi)$ *be given by formula* (5.2.1) *for absolutely continuous* ϕ, *and put* $S(\phi) = +\infty$ *for all other functions.*

Let F be a bounded functional on $\boldsymbol{D}_0[0, T]$, *measurable with respect to the σ-algebra generated by the cylinder sets, and continuous in the topology of uniform convergence. Let the maximum of the functional* $F - S$ *be attained on a unique function* $\phi_0 \in \boldsymbol{D}_0[0, T]$; *let this function belong to* $\boldsymbol{C}^1_0[0, T]$, *and let the functional F be twice Fréchet differentiable at the point* ϕ_0 *in the norm* $\|\cdot\|$.

Let

$$F''(\phi_0)(x, x) < S''(\phi_0)(x, x)$$

for any non-zero function $x \in \boldsymbol{W}^{1,2}_0[0, T]$, *where the second derivative* S'' *is extended from* $\boldsymbol{C}^1_0[0, T]$ *to* $\boldsymbol{W}^{1,2}_0[0, T]$ *by formula* (5.2.3).

Then, as $h \downarrow 0$,

$$\mathsf{M}^{h,h}_{0,0} \exp\{h^{-1} F(\xi^{h,h})\} \sim K_0 \exp\{h^{-1}[F(\phi_0) - S(\phi_0)]\}, \quad (5.2.4)$$

where

$$K_0 = \mathsf{M} \exp \{Q(\eta)\} < \infty, \tag{5.2.5}$$

$\eta(t)$, $0 \le t \le T$, $\eta(0) = 0$, is a Gaussian diffusion process in R^r with diffusion matrix

$$(A^{ij}(t)) = \left(\frac{\partial^2 G_0}{\partial z_i \partial z_j} (t, \phi_0(t);\ \nabla_u H_0(t, \phi_0(t);\ \dot{\phi}(t))) \right) \tag{5.2.6}$$

and drift coefficients

$$B^i(t, x) = \sum_{j=1}^{r} C^i_j(t)\, x^j, \tag{5.2.7}$$

$$C^i_j(t) = \frac{\partial^2 G_0}{\partial z_i \partial x^j} (t, \phi_0(t);\ \nabla_u H_0(t, \phi_0(t);\ \dot{\phi}(t))), \tag{5.2.8}$$

and the functional Q on $\mathbf{D}_0[0, T]$ is given by

$$Q(x) = \frac{1}{2} \left[F''(\phi_0)(x, x) + \int_0^T \sum_{i, j} D_{ij}(t)\, x^i(t)\, x^j(t)\, dt \right] \tag{5.2.9}$$

where

$$D_{ij}(t) = \frac{\partial^2 G_0}{\partial x^i \partial x^j} (t, \phi_0(t);\ \nabla_u H_0(t, \phi_0(t);\ \dot{\phi}_0(t))). \tag{5.2.10}$$

Note that under the conditions of Theorem 5.1.1 we have $S''(\phi_0)(x, x) = 2S(x)$, $A^{ij}(t) = \delta^{ij}$, $B^i(t, x) = 0$, $D_{ij}(t) = 0$.

Proof. First of all, according to Theorem 4.3.1, $h^{-1} S_{0,T}(\phi)$ is the action functional for the considered family of processes as $h \downarrow 0$. Just as in § 5.1, we discard the complement of the ε-neighbourhood of the point ϕ_0 and use the generalized Cramér's transformation with $z^h(t) = h^{-1} z_0(t)$, $z_0(t) = \nabla_u H_0(t, \phi_0(t);\ \dot{\phi}_0(t))$:

$$\mathsf{P}^{z^h}(A) = $$
$$= \mathsf{M}^{h, h}_{0, 0} \left(A;\ \exp \left\{ h^{-1} \int_0^T z_0(t)\, d\xi^{h, h}(t) - h^{-1} \int_0^T G_0(t, \xi^{h, h}(t);\ z_0(t))\, dt \right\} \right).$$

Using the notation M^{z^h} for the corresponding expectation and putting

$$\eta^h(t) = h^{-1/2} (\xi^{h, h}(t) - \phi_0(t)),$$

we obtain:

$$\mathsf{M}_{0,\,0}^{h,\,h}\exp\{h^{-1}F(\xi^{h,\,h})\}\sim$$

$$\sim\mathsf{M}_{0,\,0}^{h,\,h}\{\rho_{0,\,T}(\xi^{h,\,h},\phi_0)<\varepsilon;\ \ \exp\{h^{-1}F(\xi^{h,\,h})\}\}=$$

$$=\mathsf{M}^{z^h}\Bigg\{\rho_{0,\,T}(\xi^{h,\,h},\phi_0)<\varepsilon;\ \ \exp\Bigg\{h^{-1}F(\xi^{h,\,h})-$$

$$-h^{-1}\int_0^T z_0(t)\,d\xi^{h,\,h}(t)+h^{-1}\int_0^T G_0(t,\xi^{h,\,h}(t);\ \ z_0(t)\,dt\Bigg\}\Bigg\}=$$

$$=\mathsf{M}^{z^h}\Bigg\{\|\eta^h\|<h^{-1/2}\varepsilon;\ \ \exp\Bigg\{h^{-1}F(\phi_0+h^{1/2}\eta^h)-$$

$$-h^{-1/2}\int_0^T z_0(t)\,d\eta^h(t)-h^{-1}\int_0^T[z_0(t)\,\dot\phi_0(t)-$$

$$-G_0(t,\phi_0(t)+h^{1/2}\eta^h(t);\ \ z_0(t))]\,dt\Bigg\}\Bigg\}.$$

Taking Taylor expansions of the functional F near the point ϕ_0 and of the function G_0 in its second argument near the point $\phi_0(t)$ up to terms of second order, we obtain:

$$\mathsf{M}_{0,\,0}^{h,\,h}\exp\{h^{-1}F(\xi^{h,\,h})\}\sim\mathsf{M}^{z^h}\chi_{\{\|\eta^h\|<h^{-1}\varepsilon\}}\times$$

$$\times\exp\Bigg\{h^{-1}\Bigg[F(\phi_0)-\int_0^T[z_0(t)\,\dot\phi_0(t)-G_0(t,\phi_0(t);\ \ z_0(t))]\,dt\Bigg]+$$

$$+h^{-1}\Bigg[F'(\phi_0)(\eta^h)-\int_0^T z_0(t)\,d\eta^h(t)+$$

$$+\int_0^T\nabla_x G_0(t,\phi_0(t);\ \ z_0(t))\,\eta^h(t)\,dt\Bigg]+\frac{1}{2}\Bigg[F''(\phi_0)(\eta^h,\eta^h)+$$

$$+\int_0^T\sum_{i,\,j}\frac{\partial^2 G_0}{\partial x^i\partial x^j}(t,\phi_0(t);\ \ z_0(t))\,\eta^{hi}(t)\,\eta^{hj}(t)\,dt\Bigg]+$$

$$+h^{-1}R(h^{1/2}\eta^h)\Bigg\},\qquad(5.2.11)$$

where $R(x) = o(\| x \|^2)$ as $x \to 0$ ($x \in D_0[0, T]$).

5.2.3. The linear terms cancel

The factor at h^{-1} is nothing but $F(\phi_0) - S(\phi_0)$. We prove, as in § 5.1, that the terms involving $h^{-1/2}$ cancel each other with probability 1.

Since the functional $F - S$ has an extremum at the point ϕ_0 and since this functional is differentiable along the subspace $C_0^1[0, T] \subset D_0[0, T]$, we have

$$F'(\phi_0)(x) - S'(\phi_0)(x) = 0$$

for all $x \in C_0^1[0, T]$. Apply Lemmas 5.2.1 and 5.2.3 to the functional $F'(\phi_0)(x)$: the terms with $x^i(t_j) - x^i(t_j-)$ vanish:

$$\int_{(0, T]} \sum_i x^i(t) dV_i(t) - \int_0^T \sum_i \frac{\partial H_0}{\partial x^i} x^i(t)\, dt - \int_0^T \sum_i \frac{\partial H_0}{\partial u^i} \dot{x}^i(t)\, dt = 0$$

for all $x \in C_0^1[0, T]$. We standardize the functions $V_i(t)$ by imposing the condition $V_i(T) = 0$; put

$$W_i(t) = \int_t^T \frac{\partial H_0}{\partial x^i}(s, \phi_0(s); \dot{\phi}_0(s))\, ds.$$

Integrate by parts the first two integrals in the formula above; the side terms vanish because $x^i(0) = 0$, $V_i(T) = W_i(T) = 0$. We obtain:

$$\int_0^T \sum_i \left[-V_i(t) - W_i(t) - \frac{\partial H_0}{\partial u^i}(t, \phi_0(t); \dot{\phi}_0(t)) \right] \dot{x}^i(t)\, dt = 0.$$

Since $\dot{x}^i(t)$ are arbitrary continuous functions, we have for almost all t,

$$-V_i(t) - W_t(t) - \frac{\partial H_0}{\partial u^i}(t, \phi_0(t)); \dot{\phi}_0(t)) = 0. \tag{5.2.12}$$

Now we consider the terms involving the factor $h^{-1/2}$ in (5.2.11). The jumps of the locally infinitely divisible process $\xi^{h,h}(t)$, as well as those of $\eta^h(t)$, fall into the countable set $\{t_j\}$ only with probability zero; therefore the terms involving $\eta^{hi}(t_j) - \eta^{hi}(t_j-)$ in $F'(\phi_0)(\eta^h)$ can be disregarded. We obtain almost surely:

$$F'(\phi_0)(\eta^h) - \int_0^T z_0(t)\, d\eta^h(t) + \int_0^T \nabla_x G_0(t, \phi_0(t);\; z_0(t))\, \eta^h(t)\, dt =$$

$$= \int_{(0,T]} \sum_i \eta^{hi}(t)\, dV_i(t) - \int_0^T \sum_i z_{0i}(t)\, d\eta^{hi}(t) +$$

$$+ \int_0^T \sum_i \frac{\partial G_0}{\partial x^i}(t, \phi_0(t);\; z_0(t))\, \eta^{hi}(t)\, dt. \tag{5.2.13}$$

Integrate by parts the first integral, and transform it to a stochastic integral (the side term vanishes because $V_i(T) = 0$, $\eta^{hi}(0) = 0$); as for the second, we must remember that

$$z_{0i}(t) = \frac{\partial H_0}{\partial u^i}(t, \phi_0(t);\; \dot{\phi}_0(t)).$$

In the third, using Lemma 5.2.2, we substitute $-\dfrac{\partial H_0}{\partial x^i}(t, \phi_0(t);\; \dot{\phi}_0(t)) = \dfrac{dW_i(t)}{dt}$ for $\dfrac{\partial G_0}{\partial x^i}(t, \phi_0(t);\; z_0(t))$ and integrate by parts too. We obtain that the expression (5.2.13) is, with probability one, equal to

$$\int_0^T \sum_i \left[-V_i(t) - \frac{\partial H_0}{\partial u^i}(t, \phi_0(t);\; \dot{\phi}_0(t)) - W_i(t) \right] d\eta^{hi}(t),$$

i.e., according to (5.2.12) equals zero.

5.2.4. Representation as an integral over $D_0[0, T]$. Outline of the remainder of the proof

Now rewrite formula (5.2.11), taking into account that all terms of order one in η^h vanish and using the integral with respect to the measure μ_{η^h} which is the distribution of the random function η^h in the space $D_0[0, T]$ with respect to the probability measure P^{z^h}:

$$\mu_{\eta^h}(A) = \mathsf{P}^{z^h}\{\eta^h \in A\}.$$

We obtain:

$$\mathsf{M}_{0,\,0}^{h,\,h} \exp\{h^{-1} F(\xi^{h,\,h})\} \sim$$

$$\sim \exp\{h^{-1}[F(\phi_0) - S(\phi_0)]\} \int_{D_0[0,\,T]} f^h(x)\, \mu_{\eta^h}(dx), \qquad (5.2.14)$$

where

$$f^h(x) = \chi_{[0,\,h^{-1/2}\varepsilon)}(\|x\|) \cdot \exp\left\{\frac{1}{2}\left[F''(\phi_0)(x, x) + \right.\right.$$

$$\left.\left. + \int_0^T \sum_{i,j} \frac{\partial^2 G_0}{\partial x^i \partial x^j}(t, \phi_0(t);\; z_0(t))\, x^i(t)\, x^j(t)\, dt\right] + h^{-1} R(h^{1/2} x)\right\}. \qquad (5.2.15)$$

It is easily seen that the functional under the integral sign converges to the functional $\exp\{Q(x)\}$ uniformly on every bounded set. We will use the following lemma.

Lemma 5.2.4. *Let a family of finite measures μ^h on a metric space X converge weakly as $h \downarrow 0$ to a finite measure μ; let measurable functions $f^h(x)$ be bounded in every bounded set and converge to a function $f(x)$ as $h \downarrow 0$, uniformly on every bounded set. Let the limit function $f(x)$ be continuous almost everywhere with respect to the measure μ.*

Then for the convergence

$$\int_X f^h(x)\, \mu^h(dx) \to \int_X f(x)\, \mu(dx)$$

as $h \downarrow 0$ and for this limit to be finite it is sufficient that the functions $f^h(x)$ are uniformly integrable with respect to the measures μ^h.

We omit the proof of this purely analytical lemma.

Outline of the remainder of the proof. On the space $\boldsymbol{D}_0[0, T]$, along with the metric $\rho_{0,T}$ we introduce Skorohod's metric. The limit functional $\exp\{Q(x)\}$ is continuous in the topology of uniform convergence. But the convergences $y \to x$ in the sense of both metrics in $\boldsymbol{D}_0[0, T]$ coincide if $x \in \boldsymbol{C}_0[0, T]$ (see Billingsley [1], § 14). Therefore the functional $\exp\{Q(x)\}$ is continuous in Skorohod's topology at all points $x \in \boldsymbol{C}_0[0, T]$, that is, almost everywhere with respect to the distribution of any diffusion process. Thus it remains to establish the weak convergence of the measures μ_{η^h} to the distribution μ_η of the diffusion process mentioned and uniform integrabilty of $f^h(x)$ with respect to μ_{η^h}.

5.2.5. The limiting process

We will not give the complete proof of convergence of μ_{η^h} to μ_η; it follows the standard pattern. First, one verifies the weak precompactness of the family of measures $\{\mu_{\eta^h}\}$; secondly, the fact that a certain characteristic of each limit point of the family of measures $\{\mu_{\eta^h}\}$ as $h \downarrow 0$, which defines the limit measure uniquely, is the one it should be. As the device used to characterize measures in $\boldsymbol{D}_0\,[0, T]$, we take martingale problems. Let $\hat{\mathcal{C}}^{1,2}$ be the set of bounded functions $f(t, x)$, $t \in [0, T]$, $x \in R^r$, once continuously differentiable in the first argument and twice in the second, and with $\left|\frac{\partial f}{\partial t}\right|$, $\left|\frac{\partial^2 f}{\partial x^i \partial x^j}\right|$, $|x| \left|\frac{\partial f}{\partial x^i}\right|$ bounded. The distributions μ_{η^h} and μ_η are characterized by means of the compensators of $f(t, x(t))$ for all functions $f \in \hat{\mathcal{C}}^{1,2}$ with respect to these measures:

$$\widetilde{f(t, x(t))}^{\mu_{\eta^h}} = \int_0^t \mathfrak{A}_{\eta^h} f(s, x(s))\, ds \tag{5.2.16}$$

and

$$\widetilde{f(t, x(t))}^{\mu_\eta} = \int_0^t \mathfrak{A}_\eta f(s, x(s))\, ds, \tag{5.2.17}$$

where $\mathfrak{A}_{\eta^h}$ and $\mathfrak{A}_\eta$ are the corresponding compensating operators.

Let us write down these operators. The compensating operator of the original process is given by formula (4.2.3) with $\tau = h$ and $b^{\tau, h} \equiv b$, i.e.,

$$\mathfrak{A}_{\xi^{h,h}} f(t, x) = \frac{\partial f}{\partial t}(t, x) + \sum_i b^i(t, x) \frac{\partial f}{\partial x^i}(t, x) +$$
$$+ \frac{h}{2} \sum_{i,j} a^{ij}(t, x) \frac{\partial^2 f}{\partial x^i \partial x^j}(t, x) +$$
$$+ h^{-1} \int \left[f(t, x + hu) - f(t, x) - h \sum_i \frac{\partial f}{\partial x^i}(t, x)\, u^i \right] \nu_{t,x}(du).$$

The compensating operator of the same process with respect to the measure subjected to Cramér's transformation is

$$\mathfrak{A}^{z^h}_{\xi^{h,h}} f(t, x) = \frac{\partial f}{\partial t}(t, x) + \sum_i \frac{\partial G_0}{\partial z_i}(t, x; z_0(t)) \frac{\partial f}{\partial x^i}(t, x) +$$

$$+ \frac{h}{2} \sum_{i,j} a^{ij}(t, x) \frac{\partial^2 f}{\partial x^i \partial x^j}(t, x) +$$

$$+ h^{-1} \int \left[f(t, x + hu) - f(t, x) - h \sum_i \frac{\partial f}{\partial x^i}(t, x) u^i \right] e^{z_0(t) u} \nu_{t,x}(du),$$

and for the process $\eta^h(t) = h^{-1/2}(\xi^{h,h}(t) - \phi_0(t))$ with respect to P^{z^h}:

$$\mathfrak{A}_{\eta^h} f(t, x) = \frac{\partial f}{\partial t}(t, x) +$$

$$+ h^{-1/2} \sum_i \left[\frac{\partial G_0}{\partial z_i}(t, \phi_0(t) + h^{1/2} x;\; z_0(t)) - \dot{\phi}_0^i(t) \right] \frac{\partial f}{\partial x^i}(t, x) +$$

$$+ \frac{1}{2} \sum_{i,j} a^{ij}(t, \phi_0(t) + h^{1/2} x) \frac{\partial^2 f}{\partial x^i \partial x^j}(t, x) +$$

$$+ h^{-1} \int \left[f(t, x + h^{1/2} u) - f(t, x) - h^{1/2} \sum_i \frac{\partial f}{\partial x^i}(t, x) u^i \right] \times$$

$$\times e^{z_0(t) u} \nu_{t, \phi_0(t) + h^{1/2} x}(du). \qquad (5.2.18)$$

Last, for the diffusion process $\eta(t)$ we have

$$\mathfrak{A}_\eta f(t, x) = \frac{\partial f}{\partial t}(t, x) + \sum_i B^i(t, x) \frac{\partial f}{\partial x^i}(t, x) + \frac{1}{2} \sum_{i,j} A^{ij}(t) \frac{\partial^2 f}{\partial x^i \partial x^j}(t, x). \qquad (5.2.19)$$

The distribution μ_h on $D_0[0, T]$ is the unique solution of the martingale problem (5.2.17) (see Stroock, Varadhan [1], [2]).

As $h \downarrow 0$, for $f \in \hat{C}^{1,2}$ convergence of the operators takes place. Indeed, in (5.2.18) we have

$$\dot{\phi}_0^i(t) = \frac{\partial G_0}{\partial z_i}(t, \phi_0(t);\; z_0(t)).$$

Using Taylor expansion we obtain

$$\mathfrak{A}_{\eta^h} f(t, x) = \frac{\partial f}{\partial t}(t, x) + \sum_{i,j} \frac{\partial^2 G_0}{\partial z_i \partial x^j}(t, \phi_0(t) + \theta_i h^{1/2} x;\; z_0(t)) x^j \frac{\partial f}{\partial x^i}(t, x) +$$

$$+\frac{1}{2}\sum_{i,j} a^{ij}(t, \phi_0(t) + h^{1/2}x)\frac{\partial^2 f}{\partial x^i \partial x^j}(t, x) +$$

$$+\int \frac{1}{2}\sum_{i,j}\frac{\partial^2 f}{\partial x^i \partial x^j}(t, x + \theta h^{1/2} u)\, u^i u^j \nu_{t, \phi_0(t) + h^{1/2}x}(du),$$

where θ, $\theta_i \in (0, 1)$. The second term converges to

$$\sum_{i,j}\frac{\partial^2 G_0}{\partial z_i \partial x^j}(t, \phi_0(t);\ z_0(t))\, x^j \frac{\partial f}{\partial x^i}(t, x) = \sum_i B^i(t, x)\frac{\partial f}{\partial x^i}(t, x).$$

The third and fourth together yield

$$\frac{1}{2}\sum_{i,j}\Bigg[a^{ij}(t, \phi_0(t) + h^{1/2}x) +$$

$$+\int e^{z_0(t)u} u^i u^j \nu_{t, \phi_0(t) + h^{1/2}x}(du)\Bigg]\frac{\partial^2 f}{\partial x^i \partial x^j}(t, x) + o(1);$$

the sum between the brackets is

$$\frac{\partial^2 G_0}{\partial z_i \partial z_j}(t, \phi_0(t) + h^{1/2}x;\ z_0(t)) \to A^{ij}(t).$$

So, $\mathfrak{A}_{\eta^h} f(t, x) \to \mathfrak{A}_{\eta} f(t, x)$ as $h \downarrow 0$, and the convergence is uniform in every bounded set. The weak convergence of μ_{η^h} to μ_η is deduced from this fact in the same way as it is done (in another situation) by Stroock and Varadhan ([2], Chapter 11).

5.2.6. Uniform integrability. End of proof

To establish uniform integrability of the $f^h(x)$ with respect to μ_{η^h}, it is sufficient to verify that for some positive κ there exists a positive constant C such that

$$\mathsf{M}^{z^h} f^h(\eta^h)^{1+0,1\kappa} \le C.$$

In § 5.1, the functional S played the rôle of the normalized action functional for the family of processes ξ^h, of half its second derivative at an arbitrary point of $W^{1,2}[0, T]$, and of the action functional for the process w taking part in representations (5.1.10), (5.1.11). In our present case all these functionals are different. We have already given the functionals S and $S''(\phi_0)(x, x)$; let us write down the action functional for the limiting diffusion process η. We will denote it by $\tilde{I}(x) = \tilde{I}_{0,T}$:

$$\tilde{I}(x) = \tilde{I}_{0,T}(x) = \int_0^T \frac{1}{2} \sum_{i,j} \frac{\partial^2 H_0}{\partial u^i \partial u^j} \left(\dot{x}^i(t) - \sum_k \frac{\partial^2 G_0}{\partial z_i \partial x^k} x^k(t) \right) \times$$

$$\times \left(\dot{x}^j(t) - \sum_l \frac{\partial^2 G_0}{\partial z_j \partial x^l} x^l(t) \right) dt =$$

$$= \int_0^T \frac{1}{2} \sum_{i,j} \left[\frac{\partial^2 H_0}{\partial u^i \partial u^j} \dot{x}^i(t) \dot{x}^j(t) + 2 \frac{\partial^2 H_0}{\partial u^i \partial x^j} \dot{x}^i(t) x^j(t) + \right.$$

$$\left. + \left(\frac{\partial^2 H_0}{\partial x^i \partial x^j} + \frac{\partial^2 G_0}{\partial x^i \partial x^j} \right) x^i(t) x^j(t) \right] dt. \qquad (5.2.20)$$

Here the derivatives of H_0 are taken at the point $(t, \phi_0(t); \dot{\phi}_0(t))$, those of G_0 at the point $(t, \phi_0(t); z_0(t))$, and Lemma 5.2.2 is used.

Instead of Lemma 5.1.1 we use in our case

Lemma 5.2.5. *Let $Q(x)$ be a continuous quadratic functional on $\boldsymbol{D}_0[0, T]$; let $\tilde{I}(x)$ be the functional given by formula (5.2.20) for $x \in \boldsymbol{W}_0^{1,2}[0, T]$ and equal to $+\infty$ for all other $x \in \boldsymbol{D}_0[0, T]$. Let $Q(x) < \tilde{I}(x)$ for all non-zero $x \in \boldsymbol{W}_0^{1,2}[0, T]$. Then there exist positive constants K and κ such that $\| x \|^2 \le K\tilde{I}(x)$, $Q(x) \le (1 - \kappa)\tilde{I}(x)$ for all $x \in \boldsymbol{D}_0[0, T]$.*

The **proof** is the same as that of Lemma 5.1.1. ◊

Taking into account formulas (5.2.3), (5.2.9), (5.2.20) we deduce from

$$F''(\phi_0)(x, x) < S''(\phi_0)(x, x)$$

that $Q(x) < \tilde{I}(x)$. Applying Lemma 5.2.5, we find that there exists a $\kappa > 0$ such that $Q(x) \le (1 - \kappa)\tilde{I}(x)$ for all $x \in \boldsymbol{D}_0[0, T]$. We restrict the choice of $\varepsilon > 0$ by requiring that $R(x) \le 0.1 \frac{\kappa}{K} \| x \|^2$ for $\| x \| < \varepsilon$. Then to establish uniform integrability of the $f^h(\eta^h)$ it is sufficient to verify that

$$\mathsf{M}^{z^h}\{ \| \eta^h \| < h^{-1}\varepsilon;\ \exp\{(1 + 0.1\kappa) Q_1(\eta^h)\}\} \le C < \infty,$$

where $Q_1(x) = Q(x) + 0.1\frac{\kappa}{K}\|x\|^2$. The above expectation does not exceed

$$\sum_{m=0}^{\infty} \mathsf{M}^{z^h}\{\|\eta^h\| < h^{-1/2}\varepsilon,$$

$$\eta^h \in \tilde{\Phi}_0(m+1)_{+\alpha\sqrt{m+1}} \setminus \tilde{\Phi}_0(m)_{+\alpha\sqrt{m}};\ \exp\{(1+0.1\kappa)Q_1(\eta^h)\}\} \le$$

$$\le \sum_{m=0}^{\infty} \mathsf{P}^{z^h}\{\|\eta^h\| < h^{-1/2}\varepsilon,\ \eta^h \notin \tilde{\Phi}_0(m)_{+\alpha\sqrt{m}}\} \times$$

$$\times \exp\{(1+0.1\kappa)\sup\{Q_1(x):\ x \in \tilde{\Phi}_0(m+1)_{+\alpha\sqrt{m+1}}\}\},$$

where $\alpha > 0$ and $\tilde{\Phi}_0(m)_{+\alpha\sqrt{m}}$ is the $\alpha\sqrt{m}$-neighbourhood of the set

$$\tilde{\Phi}_0(m) = \{x \in \boldsymbol{D}_0[0, T]: \tilde{I}(x) \le m\}.$$

The last upper bound is estimated in the same way as in § 5.1; the probability, using Theorem 2.4.2. We indicate the elements of the construction used in this theorem: $V = (R^r)^2$; B is the set of all values $x(t)$, $0 \le t \le T$, of functions of $\tilde{\Phi}_0(m)_{+\alpha\sqrt{m}}$, intersected with $\{x: |x| < h^{-1/2}\varepsilon\}$; for δ we choose $\alpha\sqrt{m}$, $\delta' = \delta/3$; $A = m$; $t_i = iT/n$; $z(j)$ and $z\{j\}$ are taken proportional to $\sqrt{m}$, and $d(j) = md_0$. In contrast to § 5.1, we cannot take ε_1 equal to 0; this constant can be made small by choosing a small α and a large number n of intervals of the partition. The proof is completed in the same way as that of Theorem 5.1.1. ◊

5.2.7. Refinements

Naturally, the result can be carried over to the case of

$$\mathsf{M}_{0,0}^{h,h}\, G(\xi^{h,h}) \exp\{h^{-1} F(\xi^{h,h})\}.$$

However, in contrast to § 5.1 we cannot obtain as easily the refinement of this result for an $s-2$ times differentiable functional G and an s times differentiable F. This is due to the lack of general results on asymptotical expansions of $\mathsf{M}^{z^h} H(\eta^h)$ as $h \downarrow 0$ for smooth functionals H.

In the paper of Dubrovskii [2] results on asymptotical expansions were obtained for a special form of smooth functionals F:

$$F(\phi) = \int_0^T v(t, \phi(t))\, dt + w(\phi(T)),$$

and $G \equiv 1$. To this end the author used a still more general variant of Cramér's transformation, enabling him to almost reduce the functional under the expectation sign to a constant.

CHAPTER 6

ASYMPTOTICS OF THE PROBABILITY OF LARGE DEVIATIONS DUE TO LARGE JUMPS OF A MARKOV PROCESS

6.1. Conditions imposed on the family of processes. Auxiliary results

6.1.1. Conditions A - D

In this chapter, as in the previous ones, we consider two classes of strong Markov processes in R^r: locally infinitely divisible processes and τ-processes. But here we do not impose any conditions analogous to that of finiteness of exponential moments (or even of power ones), so we suppose that the compensating operator is given by the more general formula (1.3.1), rather than by (2.1.1). As we restrict ourselves to the case of the state space R^r, it will be convenient to write the compensating operator in another form, using only one coordinate system. Namely, define the function h of a real argument by $h(u) = u$ for $|u| \le 1$, $h(u) = 1$ for $u > 1$ and $h(u) = -1$ for $u < -1$. We write the compensating operator in the form

$$\mathfrak{A} f(t, x) = \frac{\partial f}{\partial t}(t, x) + \sum_i b^i(t, x) \frac{\partial f}{\partial x^i}(t, x) + \frac{1}{2} \sum_{i,j} a^{ij}(t, x) \frac{\partial^2 f}{\partial x^i \partial x^j}(t, x) +$$
$$+ \int_{R^r} \left[f(t, y) - f(t, x) - \sum_i \frac{\partial f}{\partial x^i}(t, x)\, h(y^i - x^i) \right] \lambda_{t, x}(dy), \qquad (6.1.1)$$

where we suppose that

$$\int_{R^r} [1 \wedge |y - x|^2]\, \lambda_{t, x}(dy) < \infty.$$

Every operator of the form (1.3.1) can be rewritten in this form, but with different coefficients $b^i(t, x)$.

Let $(\xi^\theta(t), \mathsf{P}^\theta_{t, x})$, $t \in [0, T]$, be a family of processes of one of the two classes mentioned, depending on a parameter θ changing over a set Θ with a filter $\theta \to$. We denote the characteristics of these processes by $\lambda^\theta_{t, x}$, $b^\theta(t, x)$, $a^{\theta, i, j}(t, x)$; in the case of processes changing only at times that are multiples of a positive number, we denote this number by $\tau(\theta)$ (i.e., $\tau(\theta)$-processes will be considered).

The conditions that we impose on the family under consideration will ensure the following. First, for sufficiently far θ the process must be close to a constant with probability almost one; secondly, the principal part of the probability of a large deviation from this constant must be formed due to paths performing one or several large jumps and remaining almost constant between these jumps; thirdly, we must be able to evaluate the asymptotics of this principal part of the probability.

Now we introduce our conditions.

A. There exists a positive function $g(\theta)$, $g(\theta) \to 0$ as $\theta \to$, and a measure $\lambda_{t,x}$ on R^r such that for any point x and any bounded continuous function f vanishing in a neighbourhood of x, for almost all t,

$$\lim_{\substack{\theta \to \\ x' \to x}} g(\theta)^{-1} \int_{R^r} f(y)\, \lambda^{\theta}_{t,y}(dy) = \int_{R^r} f(y)\, \lambda_{t,x_0}(dy).$$

Note that it is not supposed that

$$\int_{R^r} [1 \wedge |y - x|^2]\, \lambda_{t,x}(dy) < \infty$$

(i.e., the limiting measure $\lambda_{t,x}$ need not be Lévy's measure for any locally infinitely divisible process).

In Theorems 6.2.2, 6.2.2' a weaker condition will be required:

A'. For the given x_0 there exists a measure λ_{t,x_0} on R^r such that for any bounded continuous function vanishing in a neighbourhood of x_0, for almost all t,

$$\lim_{\substack{\theta \to \\ y \to x_0}} g(\theta)^{-1} \int_{R^r} f(x)\, \lambda^{\theta}_{t,x'}(dx) = \int_{R^r} f(x)\, \lambda_{t,x_0}(dx).$$

B. $\sup_{t,x} \lambda^{\theta}_{t,x}\{y : |y - x| \geq \delta\} \leq K_1(\delta)\, g(\theta)$ for all $\delta > 0$ and all sufficiently far θ, where $K_1(\delta) < \infty$.

C. There exists a number $\beta \in (0, 1]$ such that

$$\sup_{t,x} \left[\sum_{i,j} |a^{\theta,i,j}(t,x)| + \int_{R^r} [1 \wedge |y - x|^2]\, \lambda^{\theta}_{t,x}(dy) \right] \leq K_2 g(\theta)^{\beta}$$

for all sufficiently far θ, where $K_2 < \infty$.

D. $\sup_{t,x} |b^{\theta}(t,x)| \to 0$ as $\theta \to$.

Conditions **A**, **A'**, the most precise ones, are used (together with the condition $\tau(\theta) \to 0$ in the case of $\tau(\theta)$-processes) to find the asymptotics of probabilities of large deviations due to paths which have several large jumps and remain almost constant between them; the remaining conditions are used to establish the fact that other paths are negligible.

6.1.2. Examples of verification of A - D

We give some examples of families of Markov processes satisfying Conditions **A** - **D**.

a) Let $X_1, \ldots, X_n, \ldots$ be independent, identically distributed, one-dimensional random variables with distribution μ having power "tails": for $x \to \infty$,

$$\mu(x, \infty) = \mathsf{P}\{X_i > x\} = c_+ x^{-\alpha} + o(x^{-\alpha}), \tag{6.1.2}$$

$$\mu(-\infty, -x] = \mathsf{P}\{X_i \le -x\} = c_- x^{-\alpha} + o(x^{-\alpha}); \tag{6.1.3}$$

let α belong to $(0, 1)$. Consider the family of processes $(\xi^{n,z}(t), \mathsf{P}^{n,z}_{t,x})$ depending on the two-dimensional parameter $\theta = (n, z)$ (n being a natural, z a positive number): for the process starting at time 0 from the point 0,

$$\xi^{n,z}(t) = (X_1 + \ldots + X_{[nt]})/z, \quad t \in [0, 1] \tag{6.1.4}$$

(for the process starting at time t_0 from the point x_0,

$$\xi^{n,z}(t) = x_0 + z^{-1} \sum_{nt_0 < k \le nt} X_k).$$

As the filter on the set of pairs (n, z) we consider $n \to \infty$, $z/n^{1/\alpha} \to \infty$ (the latter means that the question concerns large deviations). We have (using the definition of Lévy measure in the discrete case given in § 1.3.3):

$$\tau(n, z) = n^{-1}; \tag{6.1.5}$$

$$\lambda^{n,z}_{t,x}(A) = n \cdot \mu(z(A - x)); \tag{6.1.6}$$

$$g(n, z) = nz^{-\alpha} \to 0; \tag{6.1.7}$$

$$\lambda_{t,x}(dy) = \begin{cases} c_+\alpha\,(y - x)^{-\alpha - 1}\,dy, & y > x \\ c_-\alpha\,|y - x|^{-\alpha - 1}\,dy, & y < x; \end{cases} \tag{6.1.8}$$

$$b^{n,z}(t, x) = n \int h(u/z)\,\mu(du). \tag{6.1.9}$$

Conditions **A**, **B** are verified very easily; **C**, **D** deal with the limiting behaviour of the integrals (6.1.9) and

$$n \cdot \int [1 \wedge (u/z)^2]\,\mu(du). \tag{6.1.10}$$

Both these integrals do not exceed const $nz^{-\alpha}$ for sufficiently small $nz^{-\alpha}$, i.e., Conditions **C** (with $\beta = 1$) and **D** are satisfied; the proof is the same in both cases. We carry it out for (6.1.10).

We choose an A such that $o(x^{-\alpha})$ in (6.1.2), (6.1.3) are at most $x^{-\alpha}$ for $x \ge A$; then we split the integral (6.1.10) into three parts:

$$n \int_{-\infty}^{-A} [1 \wedge (u/z)^2]\, \mu\,(du) + n \int_{-A}^{A} [1 \wedge (u/z)^2]\, \mu\,(du) +$$

$$+ n \int_{A}^{\infty} [1 \wedge (u/z)^2]\, \mu\,(du). \tag{6.1.11}$$

The middle integral does not exceed $n \cdot A^2 z^{-2} = o\,(nz^{-\alpha})$; subject the first and the last one to a transformation consisting of twice repeated integration by parts in opposite directions. Let us show how this is done for the integral from A to ∞:

$$n \int_{A}^{\infty} [1 \wedge (u/z)^2]\, d\,(-\,\mu\,(u, \infty)) =$$

$$= n \left\{ \mu\,(A, \infty) \cdot [1 \wedge (A/z)^2] + \int_{A}^{\infty} \mu\,(u, \infty)\, d\,[1 \wedge (u/z)^2] \right\} \leq$$

$$\leq n \left\{ (c_+ + 1)\, A^{-\alpha} \cdot [1 \wedge (A/z)^2] + \int_{A}^{\infty} (c_+ + 1)\, u^{-\alpha}\, d\,[1 \wedge (u/z)^2] \right\} =$$

$$= (c_+ + 1) \cdot n \int_{A}^{\infty} [1 \wedge (u/z)^2]\, d\,(-\,u^{-\alpha}) =$$

$$= (c_+ + 1) \cdot n \int_{A/z}^{\infty} [1 \wedge x^2]\, d\,(-\,(zx)^{-\alpha}) =$$

$$= (c_+ + 1) \cdot nz^{-\alpha} \int_{A/z}^{\infty} [1 \wedge x^2]\, \alpha x^{-\alpha-1} dx. \tag{6.1.12}$$

Here we used the fact that the function $1 \wedge (u/z)^2$ is non-decreasing on the right half-line.

The integral at the right-hand side has a finite limit as $z \to \infty$, and so the integral (6.1.10) does not exceed const $nz^{-\alpha}$.

When estimating the integral (6.1.9) we use the inequality

$$\int_{A/z}^{\infty} h\,(x)\, \alpha x^{-\alpha-1} dx < \int_{0}^{\infty} [1 \wedge x]\, \alpha x^{-\alpha-1} dx < \infty.$$

b) The same example as a) but with $\alpha > 2$ and the expectation $MX_i = 0$; as the filter we take $n \to \infty$, $z \geq n^{1/2+\kappa}$ (where κ is a positive constant). Here the filter does not include all large deviations (they are characterized by $n \to \infty$, $z/n^{1/2} \to \infty$) but only some part of them. Formulas (6.1.5) - (6.1.9) of Example a) and Conditions **A**, **B** remain true; Conditions **C**, **D** again deal with the integrals (6.1.9), (6.1.10). Again we represent the last integral in the form (6.1.11). The middle integral does not exceed $nA^2 z^{-2}$, which is at most const $\cdot (nz^{-\alpha})^\beta$ for

$$0 < \beta \le \frac{2\kappa}{\alpha\left(\frac{1}{2} + \kappa\right) - 1}.$$

The integral at the right-hand side of (6.1.12) tends to infinity as $z \to \infty$; it is of order $z^{\alpha - 2}$. So the integral (6.1.10) does not exceed const $\cdot\, nz^{-2} \le$ const $\cdot\, (nz^{-\alpha})^\beta$ for the values of β indicated above.

Consider the integral (6.1.9). By virtue of $\mathsf{M}X_i = 0$ it is equal to

$$n \int [h(u/z) - u/z]\, \mu(du) =$$

$$= -n \int_{-\infty}^{-z} (u/z + 1)\, \mu(du) - n \int_{z}^{\infty} (u/z - 1)\, \mu(du).$$

Each of these integrals is estimated similarly to (6.1.12); thus, the second one with the factor n does not exceed (for $z \ge A$)

$$(c_+ + 1) \cdot n \int_z^\infty (u/z - 1)\, d(-u^{-\alpha}) =$$

$$= (c_+ + 1) \cdot nz^{-\alpha} \int_1^\infty (x - 1)\, \alpha x^{-\alpha - 1} dx = \text{const} \cdot nz^{-\alpha} \to 0.$$

We can consider in a similar way the case of $1 < \alpha < 2$ too.

c) Let $(\xi(t), \mathsf{P}_{t,x})$, $0 \le t \le T$, be a locally infinitely divisible process without any parameter, with local characteristics $b(t, x)$, $a^{ij}(t, x)$, $\Lambda_{t,x}$. Let

$$|b(t, x)|, \; |a^{ij}(t, x)|, \; \int [1 \wedge |y - x|^2]\, \Lambda_{t,x}(dy) \le \text{const} < \infty;$$

suppose that for any x and for any bounded continuous function $f(y)$ vanishing in a neighbourhood of x,

$$\lim_{\substack{x' \to x \\ t \downarrow 0}} \int f(y)\, \Lambda_{t,x'}(dy) = \int f(y)\, \Lambda_{0,x}(dy)$$

(that is, the values of the measure $\Lambda_{t,x}$ away from the point x depend on the arguments t, x at $t = 0$ in a weakly continuous way).

For $\theta \in (0, 1]$, consider the Markov process $\xi^\theta(t) = \xi(\theta t)$ with time-scale changed by θ times; let the parameter θ tend to 0. Here

$$\lambda^{\theta}_{t,x}(A) = \theta\Lambda_{\theta t,x}(A),$$
$$g(\theta) = \theta \to 0,$$
$$\lambda_{t,x}(A) = \Lambda_{0,x}(A),$$
$$b^{\theta}(t,x) = \theta b(\theta t, x),$$
$$a^{\theta,i,j}(t,x) = \theta a^{ij}(\theta t, x);$$

Conditions **A** - **D** are fulfilled.

6.1.3. The times $\tau^{\varepsilon}(s)$, τ^{ε}_k. Auxiliary results

For arbitrary $\varepsilon > 0$ and $s \geq 0$ we define the first time of a jump of size $\geq \varepsilon$ after s:

$$\tau^{\varepsilon}(s) = \min\{t > s: |\xi^{\theta}(t) - \xi^{\theta}(t-)| \geq \varepsilon\}.$$

If there are no such t, put $\tau^{\varepsilon}(s) = +\infty$. Define by induction a sequence of Markov times: $\tau^{\varepsilon}_0 = 0$, $\tau^{\varepsilon}_k = \tau^{\varepsilon}(\tau^{\varepsilon}_{k-1})$ (the time of the k-th jump of size $\geq \varepsilon$); denote by ν^{ε} the number of jumps of size $\geq \varepsilon$ (i.e., the number of all $i \geq 1$ such that $\tau^{\varepsilon}_i \leq T$).

Lemma 6.1.1. *Under Condition* **B** *we have*

$$\mathsf{M}^{\theta}_{0,x_0}\, \nu^{\varepsilon}(\nu^{\varepsilon}-1)\ldots(\nu^{\varepsilon}-k+1) = O(g(\theta)^k), \quad \mathsf{P}^{\theta}_{0,x_0}\{\nu^{\varepsilon} \geq k\} = O(g(\theta)^k),$$

uniformly with respect to x_0.

Proof. We have

$$\mathsf{M}^{\theta}_{0,x_0}\, \nu^{\varepsilon}(\nu^{\varepsilon}-1)\ldots(\nu^{\varepsilon}-k+1) = k! \quad \mathsf{M}^{\theta}_{0,x_0} \sum_{0,T}^{k} (V),$$

where

$$V(t_1, y_1, x_1, \ldots, t_k, y_k, x_k) = 1$$

if all $|x_i - y_i| \geq \varepsilon$, and $V = 0$ if $|x_i - y_i| < \varepsilon$ for at least one $i = 1, \ldots, k$ (see the notation in § 1.3, proof of Lemma 1.3.2). Applying Lemma 1.3.2, we obtain (for a $\tau(\theta)$-process as well as for a locally infinitely divisible process):

$$\mathsf{M}^{\theta}_{0,x_0}\, \nu^{\varepsilon}(\nu^{\varepsilon}-1)\ldots(\nu^{\varepsilon}-k+1) \leq$$
$$\leq k! \int_0^T dt_1 \int_{t_1}^T dt_2 \ldots \int_{t_{k-1}}^T dt_k \cdot [\sup_{t,x} \lambda^{\theta}_{t,x}\{y: |y-x| \geq \varepsilon\}]^k \leq$$
$$\leq T^k \cdot K_1(\varepsilon)^k g(\theta)^k.$$

The second statement of the lemma is obtained by Chebyshev's inequality. ◊

We will also use the following variant of this lemma:

Lemma 6.1.2. *Under Condition* **B**, $\mathsf{P}^{\theta}_{t,x}\{\tau^{\varepsilon}(t) \le T\} = O(g(\theta))$ *uniformly with respect to t and x.*

Lemma 6.1.1 means that with probability $1 - O(g(\theta)^k)$ the process performs at most $(k-1)$ "large" jumps in the time interval $[0, T]$. The behaviour of the process between the jumps is described by the following lemmas.

Lemma 6.1.3. *Under Conditions* **B, C, D,** *for* $\varepsilon \le 1$ *we have, uniformly with respect to s, x:*

$$\mathsf{P}^{\theta}_{s,x}\{\sup\{|\xi^{\theta}(t) - x| : t \in [s, T] \cap [s, \tau^{\varepsilon}(s))\} \ge \varepsilon\} = O(g(\theta)^{\beta}).$$

Note that here again, as in § 2.4, we consider the semi-metric $\rho_{s, T \wedge \tau_V -}$, where $V = \{(x, y) : |x - y| < \varepsilon\}$.

Proof. Consider the stochastic process $\overset{\theta}{\xi}_V$ defined for $t \in [s, T]$ by:

$$\overset{\theta}{\xi}_V(t) = \begin{cases} \overset{\theta}{\xi}(t) & \text{for } s \le t < \tau^{\varepsilon}(s), \\ \overset{\theta}{\xi}(\tau^{\varepsilon}(s)-) & \text{for } \tau^{\varepsilon}(s) \le t \le T. \end{cases}$$

In contrast to § 2.4 where such a stochastic process was introduced, we do not perform any transformation of the measure $\mathsf{P}^{\theta}_{s,x}$. For a smooth function f the compensator of $f(\overset{\theta}{\xi}_V(t))$ with respect to the probability $\mathsf{P}^{\theta}_{s,x}$ is given by

$$\widetilde{f(\overset{\theta}{\xi}_V(t))} = f(x) + \int_s^{t \wedge \tau^{\varepsilon}(s)} A_v^{\theta, V} f(\overset{\theta}{\xi}_V(v))\, dv,$$

$$A_t^{\theta, V} f(x) = \sum_i b_V^{\theta, i}(t, x) \frac{\partial f}{\partial x^i}(x) + \frac{1}{2} \sum_{i,j} a^{\theta, i, j}(t, x) \frac{\partial^2 f}{\partial x^i \partial x^j}(x) +$$

$$+ \int_{|y-x|<\varepsilon} \left[f(y) - f(x) - \sum_i \frac{\partial f}{\partial x^i}(x)\, h(y^i - x^i) \right] \lambda^{\theta}_{t,x}(dy),$$

$$b_V^{\theta, i}(t, x) = b^{\theta, i}(t, x) - \int_{|y-x| \ge \varepsilon} h(y^i - x^i)\, \lambda^{\theta}_{t,x}(dy).$$

Under our conventions concerning notations, this formula is valid for $\tau(\theta)$-processes too, if we take s and t multiples of $\tau(\theta)$.

Hence follow, in particular, expressions for the compensators and quadratic compensators of the coordinates of the process $\overset{\theta}{\xi}_V(t)$: for $\varepsilon \le 1$,

$$\widetilde{\overset{\theta, i}{\xi}_V}(t) = x^i + \int_s^{t \wedge \tau^\varepsilon(s)} b_V^{\theta, i}(v, \overset{\theta}{\xi}_V(v))\, dv,$$

$$\langle \overset{\theta, i}{\xi}_V, \overset{\theta, i}{\xi}_V \rangle (t) = \int_s^{t \wedge \tau^\varepsilon(s)} A_V^{\theta, i, i}(v, \overset{\theta}{\xi}_V(v))\, dv,$$

$$A_V^{\theta, i, i}(t, x) = a^{\theta, i, i}(t, x) + \int_{|y - x| < \varepsilon} (y^i - x^i)^2 \lambda_{t, x}(dy).$$

We can easily deduce from Conditions **D** and **B** that

$$\sup_{t, x} |b_V^\theta(t, x)| \to 0 \text{ as } \theta \to;$$

from Condition **C**, that

$$\sup_{t, x} A_V^{\theta, i, i}(t, x) = O(g(\theta)^\beta).$$

For sufficiently far θ the inequality $T \cdot \sup_{t, x} |b^\theta(t, x)| < \varepsilon/2$ is satisfied; using Kolmogorov's inequality, we obtain:

$$\mathsf{P}_{s, x}^\theta\{\sup\{|\overset{\theta}{\xi}(t) - x| : t \in [s, T] \cap [s, \tau^\varepsilon(s))\} \ge \varepsilon\} =$$

$$= \mathsf{P}_{s, x}^\theta\left\{\sup_{t \in [s, T]} |\overset{\theta}{\xi}_V(t) - x| \ge \varepsilon\right\} \le$$

$$\le \mathsf{P}_{s, x}^\theta\left\{\sup_{t \in [s, T]} |\overset{\theta}{\xi}_V(t) - \widetilde{\overset{\theta}{\xi}_V}(t)| \ge \varepsilon/2\right\} \le$$

$$\le \frac{(T - s) \cdot \sup_{t, x} \sum_i A_V^{\theta, i, i}(t, x)}{(\varepsilon/2)^2} = O(g(\theta)^\beta). \ \Diamond$$

Lemma 6.1.4. *Under Conditions* **B**, **C**, **D**, *for any natural m we have, uniformly with respect to s, x:*

$$\mathsf{P}_{s, x}^\theta\{\sup\{|\overset{\theta}{\xi}(t) - x| : t \in [s, T] \cap [s, \tau^\varepsilon(s))\} \ge (2m - 1)\varepsilon\} =$$

$$= O(g(\theta)^{m\beta}). \tag{6.1.13}$$

Proof. Denote by $A_m(\theta)$ the least upper bound of the probability at the left-hand side over all s, x; we want to prove that $A_m(\theta) = O(g(\theta)^{m\beta})$. The proof is carried out by induction; for $m = 1$ this is the assertion of the previous lemma. Introduce the Markov time $\tau^{\varepsilon}(s, x) = \tau^{\varepsilon}(s) \wedge \min\{t \geq s: |\xi^{\theta}(t) - x| \geq \varepsilon\}$. If the event under the probability sign occurs, we have $\min\{t \geq s: |\xi^{\theta}(t) - x| \geq \varepsilon\} < \tau^{\varepsilon}(s)$. The distance of the left limit of $\xi^{\theta}(t)$ at this time from x does not exceed ε; the jump is less than ε, so $|\xi^{\theta}(\tau^{\varepsilon}(s, x)) - x| \leq 2\varepsilon$, and from the event under the probability sign follows the event

$$\{\sup\{|\xi^{\theta}(t) - x'| : t \in [s', T] \cap [s', \tau^{\varepsilon}(s'))\} \geq (2m - 3)\varepsilon\},$$

where we have put for the sake of brevity $s' = \tau^{\varepsilon}(s, x)$, $x' = \xi^{\theta}(s')$ (it is clear that $\tau^{\varepsilon}(s') = \tau^{\varepsilon}(s)$). Using the strong Markov property with respect to the time $\tau^{\varepsilon}(s, x)$ we obtain that the probability at the left-hand side of (6.1.13) does not exceed

$$\mathsf{M}^{\theta}_{s, x}\{\tau^{\varepsilon}(s, x) < \tau^{\varepsilon}(s);\ \mathsf{P}^{\theta}_{s', x'}\{\sup\{|\xi^{\theta}(t) - x'| :$$

$$t \in [s', T] \cap [s', \tau^{\varepsilon}(s'))\} \geq (2m - 3)\varepsilon\}\} \leq$$

$$\leq \mathsf{P}^{\theta}_{s, x}\{\tau^{\varepsilon}(s, x) < \tau^{\varepsilon}(s)\} \cdot A_{m-1}(\theta) \leq A_1(\theta) \cdot A_{m-1}(\theta).$$

Thus we have

$$A_m(\theta) \leq A_1(\theta) \cdot A_{m-1}(\theta),$$

and taking into account the previous lemma, Lemma 6.1.4 is proved. ◊

Lemma 6.1.5. *Under Conditions* **B, C, D**, *for any integers* $k \geq 0$, $m \geq 1$, *we have, uniformly with respect to* x_0:

$$\mathsf{P}^{\theta}_{0, x_0}\{\sup\{|\xi^{\theta}(t) - \xi^{\theta}(\tau^{\varepsilon}_k)| : t \in [0, T] \cap [\tau^{\varepsilon}_k, \tau^{\varepsilon}_{k+1})\} \geq$$

$$\geq (2m - 1)\varepsilon\} = O(g(\theta)^{m\beta}).$$

The **proof** is an application of the strong Markov property with respect to the time τ^{ε}_k, using the previous lemma. ◊

6.2. Main theorems

6.2.1. The functions $x_0 \overset{t_1}{} x_1 \overset{t_2}{} \ldots x_{k-1} \overset{t_k}{} x_k$

We proved in the previous section that for far values of θ the process $\xi^{\theta}(t)$ with probability very close to 1 differs very little from a constant on every one of the intervals $[0, T] \cap [\tau^{\varepsilon}_k, \tau^{\varepsilon}_{k+1})$. In this connection we introduce the following notations.

Let there be given times $0 = t_0 < t_1 < \dots < t_k \le T$ and points $x_0, x_1, \dots, x_k \in R^r$, where $x_i \ne x_{i-1}$, $i = 1, \dots, k$. Denote by $x_0 \,{}^{t_1} x_1 \,{}^{t_2} \dots x_{k-1} \,{}^{t_k} x_k\,(t)$ the function taking the value x_0 for $t_0 \le t < t_1$, x_1 for $t_1 \le t < t_2$, ..., and, finally, x_k for $t_k \le t \le T$. In particular, for $k = 0$ we have $x_0\,(t) \equiv x_0$.

We can deduce an immediate consequence of Lemma 6.1.5, using the notation $x_0 \,{}^{\tau_1^\varepsilon} \xi^\theta\,(\tau_1^\varepsilon) \,{}^{\tau_2^\varepsilon} \dots \xi^\theta\,(\tau_{k-1}^\varepsilon) \,{}^{\tau_k^\varepsilon} \xi^\theta\,(\tau_k^\varepsilon)\,(t)$:

$$\mathsf{P}^\theta_{0,\,x_0}\left\{\nu^\varepsilon = k,\ \rho_{0,\,T}\left(\xi^\theta, x_0 \,{}^{\tau_1^\varepsilon} \xi^\theta\,(\tau_1^\varepsilon) \dots \xi^\theta\,(\tau_{k-1}^\varepsilon) \,{}^{\tau_k^\varepsilon} \xi^\theta\,(\tau_k^\varepsilon)\right) \ge (2m-1)\,\varepsilon\right\} =$$

$$= O\,(g\,(\theta)^{m\beta}). \tag{6.2.1}$$

We obtain at once the first limit theorem of the type of the law of large numbers.

Theorem 6.2.1. *Let for the family of processes* $(\xi^\theta\,(t), \mathsf{P}^\theta_{t,\,x})$ *Conditions* **B**, **C**, **D** *of the previous section be satisfied. Then*

$$\mathsf{P}^\theta_{0,\,x_0}\,\{\rho_{0,\,T}\,(\xi^\theta, x_0) \ge \delta\} = O\,(g\,(\theta))$$

as $\theta \to$, *for any* $\delta > 0$ *and uniformly with respect to* x_0.

Proof. We have for an arbitrary $\varepsilon > 0$:

$$\mathsf{P}^\theta_{0,\,x_0}\,\{\rho_{0,\,T}\,(\xi^\theta, x_0) \ge \delta\} \le$$

$$\le \mathsf{P}^\theta_{0,\,x_0}\,\{\nu^\varepsilon \ge 1\} + \mathsf{P}^\theta_{0,\,x_0}\,\{\nu^\varepsilon = 0,\ \rho_{0,\,t}\,(\xi^\theta, x_0) \ge \delta\}.$$

Lemma 6.1.1 is applied to the first summand, to the second we apply estimate (6.2.1) (with $\varepsilon = \delta/(2m-1)$, where $m \ge \beta^{-1}$). ◊

The set of all functions $x_0 \,{}^{t_1} x_1 \,{}^{t_2} \dots x_{k-1} \,{}^{t_k} x_k\,(t)$ with given x_0 and k is denoted by $B^k_{x_0}$; in particular, $B^0_{x_0}$ consists of one function, the constant x_0.

Lemma 6.1.1 and (6.2.1) give also the following result, which will be used later:

Lemma 6.2.1.

$$\mathsf{P}^\theta_{0,\,x_0}\,\{\rho_{0,\,T}\,(\xi^\theta, B^k_{x_0}) \ge \delta\} = O\,(g\,(\theta)^{k+1})$$

for any $\delta > 0$, *uniformly with respect to* x_0.

Note that

$$\rho_{0,\,T}\,(\xi^\theta,\ B^k_{x_0}) = \rho_{0,\,T}\,(\xi^\theta,\ B^0_{x_0} \cup \dots \cup B^k_{x_0}).$$

6.2.2. The measures $\mu^k_{x_0}$. One large jump

To formulate the main results, we introduce further notations. For $x_0 \in R^r$ and a natural k let us introduce a σ-finite measure $\mu^k_{x_0}$ on the set

$$E^k_{x_0} = \{(t_1, x_1, \dots, t_k, x_k) \colon 0 < t_1 < \dots < t_k \le T,\ x_i \ne x_{i-1},\ 1 \le i \le k\},$$

defining it by the equality

$$\mu^k_{x_0} \ (dt_1\, dx_1 \dots dt_k dx_k) = dt_1 \lambda_{t_1, x_0} (dx_1) \dots dt_k\, \lambda_{t_k, x_{k-1}} (dx_k).$$

For fixed x_0 and k we denote by $X_{x_0, k}$ the following mapping of $E^k_{x_0}$ into $\mathbf{D}_{x_0}$:
$(t_1, x_1, \dots, t_k, x_k) \to x_0 \,{}^{t_1} x_1 \,{}^{t_2} \dots x_{k-1} \,{}^{t_k} x_k$. The mapping $X_{x_0, k}$ is measurable; it satisfies Lipschitz's condition with constant 1 in each of the arguments $x_1, \dots, x_k$.

Theorem 6.2.2. *Let* $(\xi^\theta (t), \mathsf{P}^\theta_{t, x})$ *be a family of locally infinitely divisible processes satisfying Conditions* **A'**, **B** - **D** *of the previous section. Let* A *be a measurable subset of* $\mathbf{D}_{x_0}$ *satisfying the conditions*

$$\rho_{0,T} (A, B^0_{x_0}) > 0, \quad \lim_{\delta \downarrow 0} \mu^1_{x_0} \ (X^{-1}_{x_0} (A_{+\delta} \setminus A_{-\delta})) = 0$$

(recall that $A_{+\delta}$ *is the* δ*-neighbourhood of the set* A*, while* $A_{-\delta}$ *is the set of all points of* A *at a distance greater than* δ *from its complement* $\overline{A}$*).*

Then for $\theta \to$,

$$\mathsf{P}^\theta_{0, x_0} \{\xi^\theta \in A\} = \mu^1_{x_0} \ (X^{-1}_{x_0, 1} (A))\, g(\theta) + o(g(\theta)). \qquad (6.2.2)$$

Proof. We have to prove for an arbitrary $\kappa > 0$ that for sufficiently far θ the probability $\mathsf{P}^\theta_{0, x_0} \{\xi^\theta \in A\}$ lies within the bounds

$$[\mu^1_{x_0} \ (X^{-1}_{x_0, 1} \ (A)) \mp \kappa]\, g(\theta).$$

For a given positive κ, choose $\delta > 0$ such that

$$\delta \le \rho_{0,T} (A, x_0) / 6$$

and

$$\mu^1_{x_0} \ (X^{-1}_{x_0, 1} \ (A_{+2\delta} \setminus A_{-2\delta})) < \kappa/3.$$

The idea of the proof is to approximate the probability $\mathsf{P}^{\theta}_{0,x_0}\{\xi^{\theta} \in A\}$ by the expectation of a functional of the form $\sum^{1}_{0,T}(V)$ (see § 1.3) and use Lemma 1.3.1. The fact that the principal part of the probability is due to paths close to step functions leaves great freedom in the choice of the function $V(t, x, y)$.

Introduce the function $H(x)$ equal to 1 for $x \geq 1$, to 0 for $x \leq 0$, and $H(x) = x$ for x between 0 and 1. Put

$$\chi^-(t, x) = H(\rho_{0,T}(x_0 \,{}^t x, \bar{A}) / \delta - 1),$$

$$\chi^+(t, x) = 1 - H(\rho_{0,T}(x_0 \,{}^t x, A) / \delta - 1).$$

These are bounded measurable functions, continuous with respect to x. The function χ^- is equal to 0 for $(t, x) \notin X^{-1}_{x_0, 1}(A_{-\delta})$ and to 1 for $(t, x) \in X^{-1}_{x_0, 1}(A_{-2\delta})$; in turn, $\chi^+(t, x) = 0$ for $(t, x) \notin X^{-1}_{x_0, 1}(A_{+2\delta})$, and 1 for $(t, x) \in X^{-1}_{x_0, 1}(A_{+\delta})$. Since $\rho_{0,T}(A, B^0_{x_0}) \geq 6\delta$, both functions $\chi^{\pm}$ vanish for $|x - x_0| \leq 3\delta$.

Define the functions

$$V^{\pm}(t, y, x) = \chi^{\pm}(t, x)\,[1 - H(|x - y|/\delta - 1)].$$

According to Condition **A'**, for almost all t we have as $y \to x_0$, $\theta \to$,

$$g(\theta)^{-1} \int \chi^{\pm}(t, x)\, \lambda^{\theta}_{t, y}(dx) \to \int \chi^{\pm}(t, x)\, \lambda_{t, x_0}(dx),$$

and the left-hand side is dominated by the constant $K_1(\delta)$. Therefore

$$g(\theta)^{-1} \int_0^T dt \int \chi^{\pm}(t, x)\, \lambda^{\theta}_{t, y}(dx) \to \int_0^T dt\, \chi^{\pm}(t, x)\, \lambda_{t, x_0}(dx). \tag{6.2.3}$$

Choose a positive $\delta' \leq \delta$ so that for $|y - x_0| < \delta'$ and sufficiently far θ the left-hand side differs from the right-hand side less than $\kappa/3$. Put $\varepsilon = \delta'/(2m - 1)$, where m is an integer, $m \geq 2\beta^{-1}$.

Split $\mathsf{P}^{\theta}_{0,x_0}\{\xi^{\theta} \in A\}$ into a sum:

$$\mathsf{P}^{\theta}_{0,x_0}\{\xi^{\theta} \in A\} = \mathsf{P}^{\theta}_{0,x_0}\{\nu^{\varepsilon} = 0,\ \xi^{\theta} \in A\} +$$

$$+ \mathsf{P}^{\theta}_{0,x_0}\{\nu^{\varepsilon} = 1,\ \rho_{0,T}\left(\xi^{\theta}, x_0 \,{}^{\tau^{\varepsilon}_1}\xi^{\theta}(\tau^{\varepsilon}_1)\right) < \delta',\ \xi^{\theta} \in A\} +$$

$$+ \mathsf{P}^{\theta}_{0,x_0}\{\nu^{\varepsilon} = 1, \rho_{0,T}\left(\xi^{\theta}, {}_{x_0}^{\tau_1^{\varepsilon}}\xi^{\theta}(\tau_1^{\varepsilon})\right) \geq \delta', \ \xi^{\theta} \in A\} +$$

$$+ \mathsf{P}^{\theta}_{0,x_0}\{\nu^{\varepsilon} \geq 2, \ \xi^{\theta} \in A\}. \tag{6.2.4}$$

The first term does not exceed

$$\mathsf{P}^{\theta}_{0,x_0}\{\nu^{\varepsilon} = 0, \rho_{0,T}(\xi^{\theta}, x_0) \geq \delta'\} = O(g(\theta)^2)$$

since $\rho_{0,T}(A, x_0) \geq 6\delta'$; the third term is estimated in just the same way. The fourth does not exceed

$$\mathsf{P}^{\theta}_{0,x_0}\{\nu^{\varepsilon} \geq 2\} = O(g(\theta)^2)$$

by Lemma 6.1.1.

The principal term is the second term. For

$$\nu^{\varepsilon} = 1, \ \rho_{0,T}(\xi^{\theta}, {}_{x_0}^{\tau_1^{\varepsilon}}\xi^{\theta}(\tau_1^{\varepsilon})) < \delta'$$

it follows from $\xi^{\theta} \in A$ that ${}_{x_0}^{\tau_1^{\varepsilon}}\xi^{\theta}(\tau_1^{\varepsilon}) \in A_{+\delta}$, whence $\chi^{+}(\tau_1^{\varepsilon}, \xi^{\theta}(\tau_1^{\varepsilon})) = 1$; and conversely, from $\chi^{-}(\tau_1^{\varepsilon}, \xi^{\theta}(\tau_1^{\varepsilon})) > 0$ it follows that ${}_{x_0}^{\tau_1^{\varepsilon}}\xi^{\theta}(\tau_1^{\varepsilon}) \in A_{-\delta}$ and $\xi^{\theta} \in A$.

Furthermore, in this case $V^{\pm}(\tau_1^{\varepsilon}, \xi^{\theta}(\tau_1^{\varepsilon}-), \ \xi^{\theta}(\tau_1^{\varepsilon})) = \chi^{\pm}(\tau_1^{\varepsilon}, \xi^{\theta}(\tau_1^{\varepsilon}))$. Therefore, the second term in (6.2.4) does not exceed

$$\mathsf{M}^{\theta}_{0,x_0}\{\nu^{\varepsilon} = 1, \rho_{0,T}\left(\xi^{\theta}, {}_{x_0}^{\tau_1^{\varepsilon}}\xi^{\theta}(\tau_1^{\varepsilon})\right) < \delta'; \ V_{+}(\tau_1^{\varepsilon}, \xi^{\theta}(\tau_1^{\varepsilon}-), \xi^{\theta}(\tau_1^{\varepsilon}))\} \tag{6.2.5}$$

and is not less than

$$\mathsf{M}^{\theta}_{0,x_0}\{\nu^{\varepsilon} = 1, \rho_{0,T}\left(\xi^{\theta}, {}_{x_0}^{\tau_1^{\varepsilon}}\xi^{\theta}(\tau_1^{\varepsilon})\right) < \delta'; \ V^{-}(\tau_1^{\varepsilon}, \xi^{\theta}(\tau_1^{\varepsilon}-), \xi^{\theta}(\tau_1^{\varepsilon}))\}. \tag{6.2.6}$$

Let us use the notation $\sum_{0,T}^{1}(V^{\pm})$ (see § 1.3, proof of Lemma 1.3.2). The expectations (6.2.5), (6.2.6) are equal to

$$\mathsf{M}^{\theta}_{0,x_0}\left\{\nu^{\varepsilon} = 1, \rho_{0,T}\left(\xi^{\theta}, {}_{x_0}^{\tau_1^{\varepsilon}}\xi^{\theta}(\tau_1^{\varepsilon})\right) < \delta'; \ \sum_{0,T}^{1}(V^{\pm})\right\} =$$

$$= \mathsf{M}^{\theta}_{0,x_0}\sum_{0,T}^{1}(V^{\pm}) - \mathsf{M}^{\theta}_{0,x_0}\left\{\nu^{\varepsilon} = 1, \rho_{0,T}\left(\xi^{\theta}, {}_{x_0}^{\tau_1^{\varepsilon}}\xi^{\theta}(\tau_1^{\varepsilon})\right) \geq \delta'; \ \sum_{0,T}^{1}(V^{\pm})\right\} +$$

$$+ \mathsf{M}^{\theta}_{0,\,x_0}\left\{\nu^{\varepsilon} \geq 2;\ \sum\nolimits^{1}_{0,\,T} (V^{\pm})\right\} \tag{6.2.7}$$

(we have $\sum^{1}_{0,\,T} (V^{\pm}) = 0$ for $\nu^{\varepsilon} = 0$). The second and third expectations at the right-hand side do not exceed

$$\mathsf{P}^{\theta}_{0,\,x_0}\left\{\nu^{\varepsilon} = 1,\ \rho_{0,\,T}\left(\xi^{\theta},\ x_0{}^{\tau^{\varepsilon}_1}\xi^{\theta}\,(\tau^{\varepsilon}_1)\right) \geq \delta'\right\} = O\,(g\,(\theta)^2),$$

$$\mathsf{M}^{\theta}_{0,\,x_0}\,\{\nu^{\varepsilon} \geq 2;\ \nu^{\varepsilon}\} \leq \mathsf{M}^{\theta}_{0,\,x_0}\,\nu^{\varepsilon}\,(\nu^{\varepsilon} - 1) = O\,(g\,(\theta)^2)$$

by Lemma 6.1.1.

The first term in (6.2.7) is equal to

$$\mathsf{M}^{\theta}_{0,\,x_0}\int_0^T dt \int_{R^r} \lambda_{t,\,\xi^{\theta}(t)}\,(dx)\,V^{\pm}\,(t,\,\xi^{\theta}\,(t),\,x) \tag{6.2.8}$$

by Lemma 1.3.1. We split this expectation, in turn, into three parts:

$$\mathsf{M}^{\theta}_{0,\,x_0}\left\{\nu^{\varepsilon} = 0,\ \rho_{0,\,T}\,(\xi^{\theta},\,x_0) < \delta';\ \int_0^T\right\} +$$

$$+ \mathsf{M}^{\theta}_{0,\,x_0}\left\{\nu^{\varepsilon} = 0,\ \rho_{0,\,T}\,(\xi^{\theta},\,x_0) \geq \delta';\ \int_0^T\right\} + \mathsf{M}^{\theta}_{0,\,x_0}\left\{\nu^{\varepsilon} \geq 1;\ \int_0^T\right\}, \tag{6.2.9}$$

where the integral from 0 to T is the same as under the expectation sign in (6.2.8). This integral is $O\,(g\,(\theta))$, therefore the second and the third expectations are at most $O\,(g\,(\theta))$ multiplied by the corresponding probabilities, i.e., $O\,(g\,(\theta)^2)$.

We know that $\chi^{\pm}\,(t,\,x) > 0$ implies $|\,x - x_0\,| > 3\delta$; so $|\,\xi^{\theta}\,(t) - x_0\,| < \delta' \leq \delta$ implies that $V^{\pm}\,(t,\,\xi^{\theta}\,(t),\,x) = \chi^{\pm}\,(t,\,x)$. The integral from 0 to T in the first expectation in (6.2.9) is estimated from above and from below by the expressions

$$g\,(\theta)\left[\int_0^T dt \int_{R^r} \lambda_{t,\,x_0}\,(dx)\,\chi^{+}\,(t,\,x) + \kappa/3\right] \leq$$

$$\leq g\,(\theta)\,[\mu^1_{x_0}\,(X^{-1}_{x_0}\,(A_{+2\delta})) + \kappa/3] \leq g\,(\theta)\,[\mu^1_{x_0}\,(X^{-1}_{x_0,\,1}\,(A)) + 2\kappa/3]$$

and

$$g(\theta)\left[\int_0^T dt \int_{R^r} \lambda_{t,x_0}(dx)\,\chi^-(t,x) - \kappa/3\right] \geq$$

$$\geq g(\theta)\,[\mu^1_{x_0}(X^{-1}_{x_0}(A_{-2\delta})) - \kappa/3] \geq g(\theta)\,[\mu^1_{x_0}(X^{-1}_{x_0,1}(A)) - 2\kappa/3].$$

Thus, the first term in (6.2.8) lies within the bounds

$$g(\theta)\,(\mu^1_{x_0}(X^{-1}_{x_0,1}(A)) \mp 2\kappa/3) \cdot \mathsf{P}^\theta_{0,x_0}\{\nu^\varepsilon = 0,\ \rho_{0,T}(\xi^\theta, x_0) < \delta'\} =$$

$$= g(\theta)\,(\mu^1_{x_0}(X^{-1}_{x_0,1}(A)) \mp 2\kappa/3) + O(g(\theta)^2).$$

Finally we obtain that $\mathsf{P}^\theta_{0,x_0}\{\xi^\theta \in A\}$ is, for sufficiently far θ, within the bounds

$$g(\theta) \cdot \mu^1_{x_0}(X^{-1}_{x_0,1}(A)) \mp [g(\theta) \cdot 2\kappa/3 + O(g(\theta)^2)],$$

i.e., it differs from $\mu^1_{x_0}(X^{-1}_{x_0,1}(A)) \cdot g(\theta)$ less than $\kappa \cdot g(\theta)$. This proves the theorem. ◊

6.2.3. Discrete time

Theorem 6.2.2'. *Let $(\xi^\theta(t), \mathsf{P}^\theta_{t,x})$ be a family of $\tau(\theta)$-processes satisfying Conditions* **A'**, **B** - **D** *of the previous section; let $\tau(\theta) \to 0$ as $\theta \to$. Let A be a measurable subset of* $\mathbf{D}_{x_0}$ *satisfying the conditions*

$$\rho_{0,T}(A, B^0_{x_0}) > 0,\quad \lim_{\delta \downarrow 0} \mu^1_{x_0}([X^{-1}_{x_0,1}(A_{+\delta})] \setminus (X^{-1}_{x_0,1}(A_{-\delta}))) = 0$$

(the brackets denote the closure, the parentheses the interior of a set).

Then as $\theta \to$,

$$\mathsf{P}^\theta_{0,x_0}\{\xi^\theta \in A\} = \mu^1_{x_0}(X^{-1}_{x_0,1}(A))\,g(\theta) + o(g(\theta)).$$

The **proof** is carried out similarly to that of the previous theorem. Choose $\delta > 0$ so that $\delta < \rho_{0,T}(A, x_0)/6$ and

$$\mu^1_{x_0}([X^{-1}_{x_0,1}(A_{+\delta})] \setminus (X^{-1}_{x_0,1}(A_{-\delta}))) < \kappa/4. \tag{6.2.10}$$

Define the uniformly continuous functions

$$\chi^-(t,x) = H(\rho((t,x),\ X^{-1}_{x_0,1}(\overline{A_{-\delta}}))/\delta' - 1),$$

$$\chi^+(t,x) = 1 - H(\rho((t,x),\ X^{-1}_{x_0,1}(A_{+\delta}))/\delta' - 1).$$

A positive $\delta' \le \delta$ is chosen in such a way that

$$\iint [\chi^+(t,x) - \chi^-(t,x)]\, \mu^1_{x_0}(dt\, dx) \le \kappa/2.$$

Such a δ' exists because the limit of this integral as $\delta' \downarrow 0$ is equal to the left-hand side of (6.2.10).

Let $\gamma > 0$ be such that $|\chi^{\pm}(s,x) - \chi^{\pm}(s',x)| < \kappa/8TK_1(\delta)$ for $|s - s'| < \gamma$.

We define the functions $V^{\pm}$ as before. The only change introduced into the proof is that the integral in (6.2.8) is replaced by the sum

$$\sum_{0 \le t = k\tau(\theta) < T} \tau(\theta) \cdot \int_{R^r} \lambda^{\theta}_{t, \xi^{\theta}(t)}(dx)\, V^{\pm}(t + \tau(\theta), \xi^{\theta}(t), x) \quad (6.2.11)$$

(see formula (1.3.3)). We have to take into account that $\lambda^{\theta}_{s,\xi^{\theta}(s)}$ coincides with $\lambda^{\theta}_{k\tau(\theta), \xi^{\theta}(k\tau(\theta))}$ for $k\tau(\theta) \le s < (k+1)\tau(\theta)$. For all θ with $\tau(\theta) < \gamma$ the sum differs from

$$\int_0^T ds \int_{R^r} \lambda^{\theta}_{s, \xi^{\theta}(s)}(dx)\, V^{\pm}(s, \xi^{\theta}(s), x)$$

by at most $g(\theta) \cdot \kappa/4$ (we have $|\chi^{\pm}(s,x) - \chi^{\pm}(t + \tau(\theta), x)| < \kappa/8TK_1(\delta)$ for $t = k\tau(\theta) \le s < t + \tau(\theta)$; and we also have to take into account the integral from T to $(k_0 + 1)\tau(\theta)$, where $k_0\tau(\theta) < T \le (k_0 + 1)\tau(\theta)$). The remaining part of the proof is reproduced without changes. ◊

6.2.4. The case of k jumps

Theorem 6.2.3. *Let $(\xi^{\theta}(t), \mathrm{P}^{\theta}_{t,x})$ be a family of processes satisfying Conditions* **A** - **D** *of the previous section; in the case of $\tau(\theta)$-processes it is supposed that $\tau(\theta) \to 0$ as $\theta \to$. Let A be a measuarable subset of D_{x_0} satisfying the conditions*

$$\rho_{0,T}(A, B^{k-1}_{x_0}) > 0, \quad \lim_{\delta \downarrow 0} \mu^k_{x_0}(X^{-1}_{x_0,k}(A_{+\delta} \setminus A_{-\delta})) = 0$$

or, in the case of $\tau(\theta)$-processes,

$$\lim_{\delta \downarrow 0} \mu^k_{x_0}([X^{-1}_{x_0,k}(A_{+\delta})] \setminus (X^{-1}_{x_0,k}(A_{-\delta}))) = 0.$$

Then we have, as $\theta \to$,

$$\mathsf{P}^{\theta}_{0,\,x_0}\{\xi^{\theta} \in A\} = \mu^{k}_{x_0}(X^{-1}_{x_0,\,k}(A)) \cdot g(\theta)^{k} + o(g(\theta)^{k}). \qquad (6.2.12)$$

Proof. For simplicity we will give the proof in the case $k = 2$ and for locally infinitely divisible processes $(\xi^{\theta}(t), \mathsf{P}^{\theta}_{t,\,x})$. Let $\kappa > 0$ be given. First of all we choose the functions $\chi^{\pm}(t_1, x_1, t_2, x_2)$ in a way similar to that in the proofs of Theorems 6.2.2, 6.2.2': uniformly continuous with respect to (x_1, x_2) (in the case of $\tau(\theta)$-processes with respect to all arguments) approximations of the indicator $\chi_A(x_0{}^{t_1}x_1{}^{t_2}x_2))$, vanishing for $\rho(x_0{}^{t_1}x_1{}^{t_2}x_2, A) \geq 2\delta$. Further, it follows automatically from the fact that Condition **A** is satisfied, that it is satisfied uniformly in a certain sense and that the limit is continuous in the space variable. Namely, the following lemma is true:

Lemma 6.2.2. *Let $\mathfrak{F}$ be a uniformly bounded set, consisting of real-valued functions on R^r that is equicontinuous in every bounded region. Then for any positive K, δ and almost all t,*

$$\lim_{\substack{\delta' \downarrow 0 \\ \theta \to}} \sup \left\{ \left| \int_{R^r} g(\theta)^{-1} \lambda^{\theta}_{t,\,y'}(dx) f(x) - \int_{R^r} \lambda_{t,\,y}(dx) f(x) \right| \right\} = 0 \qquad (6.2.13)$$

and

$$\lim_{\delta' \downarrow 0} \sup \left\{ \left| \int_{R^r} \lambda_{t,\,y'}(dx) f(x) - \int_{R^r} \lambda_{t,\,y}(dx) f(x) \right| \right\} = 0, \qquad (6.2.14)$$

where the supremum is taken over all $|y| \leq K$, $|y' - y| \leq \delta'$ and over all functions $f \in \mathfrak{F}$ vanishing in the δ-neighbourhood of the point y.

We will not give the proof.

Using this lemma, choose C so that for sufficiently far θ for all y, $|y - x_0| \leq 2\delta$, we have

$$\int_0^T dt \int_{R^r} g(\theta)^{-1} \lambda^{\theta}_{t,\,y}(dx)\, H(|x| - C) \leq \kappa/10TK_1(\delta).$$

To choose such a C, we take for $\mathfrak{F}$ the set of all functions of the form $f(x) = H(|x| - C)$. Choose at first a positive δ' such that the least upper bound in (6.2.13) is at most $\kappa/20TK_1(\delta)$ for sufficiently far θ. Then in the ball $\{y: |y - x_0| \leq 2\delta\}$ we choose a finite δ'-net $y_1, \ldots, y_N$. Finally we choose $C \geq |x_0| + 3\delta + 1$ in such a way that

$$\int_0^T dt \int_{R^r} \lambda_{t, y_i} (dx)\, H\,(\,|\,x\,|-C) < \kappa/20TK_1(\delta)$$

for $i = 1, \ldots, N$.

Now, the choice of C being made, we take for $\mathfrak{F}$ the set of all functions $\chi^{\pm}$, considered as functions of the last arguments for fixed first three arguments t_1, x_1, t_2, $0 < t_1 < t_2 \le T$, $|\,x_1\,| \le C + 1$. Choose δ_1, $0 < \delta_1 \le \delta$, such that for sufficiently far θ and for all t_1, x_1, y_2, $|\,x_1\,| \le C + 1$, $|\,y_2 - x_1\,| \le 2\delta_1$, we have

$$\int_{t_1}^T dt_2 \left| \int_{R^r} g\,(\theta)^{-1} \lambda^{\theta}_{t_2, y_2} (dx_2)\, \chi^{\pm} (t_1, x_1, t_2, x_2) + \right.$$

$$\left. + \int_{R^r} \lambda_{t_2, x_1} (dx_2)\, \chi^{\pm} (t_1, x_1, t_2, x_2) \right| \le \kappa/10TK_1(\delta).$$

If we multiply $\chi^{\pm} (t_1, x_1, t_2, x_2)$ by the function

$$[1 - H\,(\,|\,x_1\,| - C)]\,[1 - H\,(\,|\,y_2 - x_1\,|/\delta_1 - 1)],$$

the corresponding inequality will hold for *all* t_1, x_1, y_2.

Now we take for $\mathfrak{F}$ the set of all functions

$$f^{\pm}_{t_1, y_2} (x_1) = [1 - H\,(\,|\,x_1\,| - C)]\,[1 - H\,(\,|\,y_2 - x_1\,|/\delta_1 - 1)] \times$$

$$\times \int_{t_1}^T dt_2 \int_{R^r} \lambda_{t_2, x_1} (dx_2)\, \chi^{\pm} (t_1, x_1, t_2, x_2)$$

for all possible t_1, y_2. All such functions are bounded by one and the same constant $T \cdot K_1(\delta)$ and are equicontinuous with respect to x_1. We choose δ_0, $0 < \delta_0 \le \delta_1 \le \delta$, so that for sufficiently far θ, for all y_1, $|\,y_1 - x_0\,| \le \delta_0$, and all y_2 we have

$$\int_0^T dt_1 \left| \int_{R^r} g\,(\theta)^{-1} \lambda^{\theta}_{t_1, y_1} (dx_1)\, f^{\pm}_{t_1, y_2} (x_1) + \right.$$

$$\left. + \int_{R^r} \lambda_{t_1, x_0} (dx_1)\, f^{\pm}_{t_1, y_2} (x_1) \right| \le \kappa/10.$$

Define the functions

$$V^{\pm} (t_1, y_1, x_1, t_2, y_2, x_2) = [1 - H\,(\,|\,y_1 - x_0\,|/\delta_0 - 1)] \times$$

$$\times [1 - H(|y_2 - x_1|/\delta_1 - 1)]\,\chi^{\pm}(t_1, x_1, t_2, x_2)$$

and

$$V_C^{\pm}(t_1, y_1, x_1, t_2, y_2, x_2) =$$
$$= [1 - H(|x_1| - C)]\, V^{\pm}(t_1, y_1, x_1, t_2, y_2, x_2).$$

Now we take a positive $\varepsilon \le \delta_0 / (2m - 1)$, where m is a natural number, $m \ge 3\beta^{-1}$; introduce the times τ_1^ε, τ_2^ε, ... and the variable ν^ε, equal to the number of indices $i \ge 1$ such that $\tau_i^\varepsilon \le T$. If $\nu^\varepsilon = 0$ and $\rho_{0,T}(\xi^\theta, x_0) < \delta_0$, or $\nu^\varepsilon = 1$ and $\rho_{0,T}(\xi^\theta, x_0 \,{}^{\tau_1^\varepsilon}\xi^\theta(\tau_1^\varepsilon)) < \delta_0$, the event $\xi^\theta \in A$ cannot occur. The probability of this event can be split into the following sum:

$$\mathsf{P}^\theta_{0,x_0}\{\xi^\theta \in A\} = \mathsf{P}^\theta_{0,x_0}\{\nu^\varepsilon = 0,\, \rho_{0,T}(\xi^\theta, x_0) \ge \delta_0,\ \xi^\theta \in A\} +$$
$$+ \mathsf{P}^\theta_{0,x_0}\left\{\nu^\varepsilon = 1,\, \rho_{0,T}\left(\xi^\theta, x_0 \,{}^{\tau_1^\varepsilon}\xi^\theta(\tau_1^\varepsilon)\right) \ge \delta_0,\ \xi^\theta \in A\right\} +$$
$$+ \mathsf{P}^\theta_{0,x_0}\left\{\nu^\varepsilon = 2,\, \rho_{0,T}\left(\xi^\theta, x_0 \,{}^{\tau_1^\varepsilon}\xi^\theta(\tau_1^\varepsilon) \,{}^{\tau_2^\varepsilon}\xi^\theta(\tau_2^\varepsilon)\right) \ge \delta_0,\ \xi^\theta \in A\right\} +$$
$$+ \mathsf{P}^\theta_{0,x_0}\left\{\nu^\varepsilon = 2,\, \rho_{0,T}\left(\xi^\theta, x_0 \,{}^{\tau_1^\varepsilon}\xi^\theta(\tau_1^\varepsilon) \,{}^{\tau_2^\varepsilon}\xi^\theta(\tau_2^\varepsilon)\right) < \delta_0,\ \xi^\theta \in A\right\} +$$
$$+ \mathsf{P}^\theta_{0,x_0}\{\nu^\varepsilon \ge 3,\ \xi^\theta \in A\}.$$

By virtue of Lemmas 6.1.1 and 6.1.4 the first, second, third and fifth terms are $O(g(\theta)^3)$; the main term is the fourth term. It is estimated from below and from above by the expectations

$$\mathsf{M}^\theta_{0,x_0}\left\{\nu^\varepsilon = 2,\, \rho_{0,T}\left(\xi^\theta, x_0 \,{}^{\tau_1^\varepsilon}\xi^\theta(\tau_1^\varepsilon) \,{}^{\tau_2^\varepsilon}\xi^\theta(\tau_2^\varepsilon)\right) < \delta_0;\right.$$
$$\left.\chi^{\pm}(\tau_1^\varepsilon, \xi^\theta(\tau_1^\varepsilon), \tau_2^\varepsilon, \xi^2(\tau_2^\varepsilon))\right\}$$
$$= \mathsf{M}^\theta_{0,x_0}\left\{\nu^\varepsilon = 2,\, \rho_{0,T}\left(\xi^\theta, x_0 \,{}^{\tau_1^\varepsilon}\xi^\theta(\tau_1^\varepsilon) \,{}^{\tau_2^\varepsilon}\xi^\theta(\tau_2^\varepsilon)\right) < \delta_0;\ \sum\nolimits^2_{0,T}(V^{\pm})\right\}. \tag{6.2.15}$$

The expression (6.2.15) differs only by $O(g(\theta)^3)$ from

$$\mathsf{M}^{\theta}_{0,x_0}\sum\nolimits^{2}_{0,T}(V^{\pm}) = \mathsf{M}^{\theta}_{0,x_0}\int_0^T dt_1\left[\int_{R^r}\lambda^{\theta}_{t_1,y_1}(dx_1)\times\right.$$

$$\left.\times\,\mathsf{M}^{\theta}_{t_1,x_1}\int_{t_1}^T dt_2\int_{R^r}\lambda^{\theta}_{t_2,\xi^{\theta}(t_2)}(dx_2)V^{\pm}(t_1,y_1,x_1,t_2,\xi^{\theta}(t_2),x_2)\right]_{y_1=\xi^{\theta}(t_1)} \quad .(6.2.16)$$

In the case of a $\tau(\theta)$-process the integrals are replaced by the corresponding sums:

$$\mathsf{M}^{\theta}_{0,x_0}\sum\nolimits^{2}_{0,T}(V^{\pm}) =$$

$$= \mathsf{M}^{\theta}_{0,x_0}\sum_{0\le t_1=k_1\tau(\theta)<T}\tau(\theta)\cdot\left[\int_{R^r}\lambda^{\theta}_{t_1,y_1}(dx_1)\times\right.$$

$$\times\,\mathsf{M}^{\theta}_{t_1+\tau(\theta),x_1}\sum_{(t_1+\tau(\theta)\le t_2=k_2\tau(\theta)<T}\tau(\theta)\,.\int_{R^r}\lambda^{\theta}_{t_2,\xi^{\theta}(t_2)}(dx_2)\times$$

$$\left.\times\,V^{\pm}(t_1,y_1,x_1,t_2,\xi^{\theta}(t_2),x_2)\right]_{y_1=\xi^{\theta}(t_1)}.$$

The integral from t_1 to T under the sign $\mathsf{M}^{\theta}_{t_1,x_1}$ in (6.2.16) does not exceed $g(\theta)\times(T-t_1)\cdot K_1(\delta)\le g(\theta)\,TK_1(\delta)$ for sufficiently far θ; replacing this expectation by

$$\mathsf{M}^{\theta}_{t_1,x_1}\left\{\sup_{t_1\le s\le T}|\xi^{\theta}(s)-x_1|<\delta_0;\right.$$

$$\left.\int_{t_1}^T dt_2\int_{R^r}\lambda^{\theta}_{t_2,\xi^{\theta}(t_2)}(dx_2)\,V^{\pm}(t_1,y_1,x_1,t_2,\xi^{\theta}(t_2),x_2)\right\},$$

we change it, by Lemma 6.1.2, only by $O(g(\theta)^2)$, and the whole expectation (6.2.16) by $O(g(\theta)^3)$. Restricting the integration range in $\mathsf{M}^{\theta}_{0,x_0}$ to

$$\left\{\sup_{0\le s\le T}|\xi^{\theta}(s)-x_0|<\delta_0\right\}$$

has the same effect.

So, $\mathsf{P}^{\theta}_{0,x_0}\{\xi^\theta \in A\}$ lies within the bounds

$$\mathsf{M}^{\theta}_{0,x_0}\left\{\sup_{0\le s\le T}|\xi^\theta(s)-x_0|<\delta_0;\ \int_0^T dt_1\left[\int_{R^r}\lambda^{\theta}_{t_1,y_1}(dx_1)\times\right.\right.$$

$$\times\,\mathsf{M}^{\theta}_{t_1,x_1}\left\{\sup_{t_1\le s\le T}|\xi^\theta(s)-x_1|<\delta_0;\ \int_{t_1}^T dt_2\int_{R^r}\lambda^{\theta}_{t_2,\xi^\theta(t_2)}(dx_2)\times\right.$$

$$\left.\left.\left.\times\, V^{\pm}(t_1,y_1,x_1,t_2,\xi^\theta(t_2),x_2)\right\}\right]_{y_1=\xi^\theta(t_1)}\right\}+O(g(\theta)^3).$$

Replacing here $V^{\pm}$ by $V^{\pm}_C$, the expectation changes by at most

$$\mathsf{M}^{\theta}_{0,x_0}\left\{\sup_{0\le s\le T}|\xi^\theta(s)-x_0|<\delta_0;\ \int_0^T dt_1\int_{R^r}\lambda^{\theta}_{t_1,\xi^\theta(t_1)}(dx_1)\times\right.$$

$$\left.\times\, H(|x_1|-C)\,.\,(T-t_1)\,g(\theta)\,K_1(\delta)\right\}\le 0.1\kappa g(\theta)^2$$

for sufficiently far θ (the latter inequality holds because of the choice of C). Then, using the choice of δ_1, δ_0, we replace $\lambda^{\theta}_{t_2,\xi^\theta(t_2)}(dx_2)$ by $g(\theta)\cdot\lambda_{t_2,x_1}(dx_2)$ and $\lambda^{\theta}_{t_1,y_1}(dx_1)$ by $g(\theta)\cdot\lambda_{t_1,x_0}(dx_1)$, introducing every time an additional term not exceeding $0.1\kappa g(\theta)^2$. Since

$$|\xi^\theta(t_1)-x_0|,\ |\xi^\theta(t_2)-x_1|<\delta_0\le\delta$$

in the integration ranges, the integrand $V^{\pm}_C$ coincides with

$$[1-H(|x_1|-C)]\cdot\chi^{\pm}(t_1,x_1,t_2,x_2).$$

Replacing this function by $\chi^{\pm}(t_1,x_1,t_2,x_2)$, we introduce once more an error that does not exceed $0.1\kappa g(\theta)^2$.

Thus, $\mathsf{P}^{\theta}_{0,x_0}\{\xi^\theta \in A\}$ lies within the bounds

$$g(\theta)^2 \cdot \int_0^T dt_1 \int_{R^r} \lambda_{t_1, x_0}(dx_1) \int_{t_1}^T dt_2 \int_{R^r} \lambda_{t_2, x_1}(dx_2) \times$$

$$\times \chi^{\pm}(t_1, x_1, t_2, x_2)\ \mathsf{P}^{\theta}_{0, x_0}\left\{\sup_{0 \le s \le T} |\xi^{\theta}(s) - x_0| < \delta_0\right\} \times$$

$$\times \mathsf{P}^{\theta}_{t_1, x_1}\left\{\sup_{t_1 \le s \le T} |\xi^{\theta}(s) - x_1| < \delta_0\right\} \mp 0.4\kappa g(\theta)^2 + O(g(\theta)^3)$$

for sufficiently far θ. Replacing the probabilities by 1, we introduce once more an error of order $O(g(\theta)^3)$.

Finally we obtain that for sufficiently far θ the probability $\mathsf{P}^{\theta}_{0, x_0}\{\xi^{\theta} \in A\}$ lies within the bounds $g(\theta)^2 [\mu^2_{x_0}(X^{-1}_{x_0, 2}(A)) \mp \kappa]$. ◊

6.2.5. Possible generalizations

In the paper of Godovan'chuk [2] more general variants of the theorems in this section are given. The principal generalization is that instead of Condition **D** (of convergence of $b^{\theta}(t, x)$ to zero) one can require that these functions converge to some function $b(t, x)$. In this case the process $\xi^{\theta}(t)$, with overwhelming probability, moves near solutions of the equation $\dot{x}(t) = b(t, x(t))$ instead of standing almost still. So in this case the function $x_0 \overset{t_1}{} x_1 \overset{t_2}{} x_2 \dots \overset{t_k}{} x_k$ is defined for $t \in [t_i, t_{i+1})$ (for $i = k$, in the interval $[t_k, T]$) not as the constant x_i but as the solution of $\dot{x}(t) = b(t, x(t))$ with initial condition $x(t_i) = x_i$.

6.3. Applications to sums of independent random variables

6.3.1. Theorem 6.3.1

Let us consider some applications of the theorems of the previous section to theorems on large deviations for independent random variables with power "tails". Using them we can obtain results on the asymptotics of probabilities of large deviations up to equivalence and, under more restrictive conditions, with refining terms. We restrict ourselves to the case of one-sided power "tails" with exponent $\alpha \in (0, 1)$. In Vinogradov [1] refined theorems on large deviations are obtained for two-sided power "tails", both for $\alpha \in (0, 1) \cup (1, 2)$, i.e. for variables attracted to a non-normal stable law, and for $\alpha > 2$.

The following result is contained in Tkachuk [1] (under more restrictive conditions, see Fortus [1], Heyde [1]); but we shall give its proof using the results of § 6.2.

Theorem 6.3.1. *Let $X_1, X_2, \ldots$ be independent identically distributed random variables, let $X_i \geq 0$, and let the distribution F of these random variables have the following asymptotics as $x \to +\infty$:*

$$F(x) = 1 - c_\alpha x^{-\alpha} + o(x^{-\alpha}), \tag{6.3.1}$$

$0 < \alpha < 1$. *Then as $n \to \infty$, $z/n^{1/\alpha} \to \infty$,*

$$\mathsf{P}\{X_1 + \ldots + X_n > z\} \sim nc_\alpha z^{-\alpha}. \tag{6.3.2}$$

Proof. We have verified (in § 6.1) that for the family of random functions

$$\xi^{n,z}(t) = (X_1 + \ldots + X_{[nt]})/z,$$

as $n \to \infty$, $z/n^{1/\alpha} \to \infty$, Conditions **A** - **D** are satisfied with $g(n, z) = nz^{-\alpha}$,

$$\lambda_{t,y}(dx) = \begin{cases} 0, & x \leq y, \\ c_\alpha d(-(x-y)^{-\alpha}), & x > y. \end{cases} \tag{6.3.3}$$

Consider the set $A = \{x \in \boldsymbol{D}_0 : x(1) > 1\}$. We have $\rho_{0,1}(A, B_0^0) = 1 > 0$; $A_{+\delta} = \{x \in \boldsymbol{D}_0 : x(1) > 1 - \delta\}$, $A_{-\delta} = \{x \in \boldsymbol{D}_0 : x(1) > 1 + \delta\}$; $X_{0,1}^{-1}(A) = (0, 1] \times (1, \infty)$, $[X_{0,1}^{-1}(A_{+\delta})] \setminus (X_{0,1}^{-1}(A_{-\delta})) = [0, 1] \times [1 - \delta, 1 + \delta] \cup \{0, 1\} \times (1 + \delta, \infty)$; the corresponding μ_0^1-measures are equal to c_α and to $c_\alpha[(1-\delta)^{-\alpha} - (1+\delta)^{-\alpha}]$ ($\to 0$ as $\delta \downarrow 0$). Now we use Theorem 6.2.2':

$$\mathsf{P}\{X_1 + \ldots + X_n > z\} = \mathsf{P}\{\xi^{n,z} \in A\} = nc_\alpha z^{-\alpha} + o(nz^{-\alpha}). \ \Diamond$$

6.3.2. Theorem 6.3.2. The terms P_0 - P_3. Estimation of P_0, P_3. Asymptotics of P_2

Theorem 6.3.2. *Let $X_1, X_2, \ldots$ be non-negative independent random variables with distribution function F,*

$$F(x) = 1 - c_{\alpha_1} x^{-\alpha_1} - c_{\alpha_2} x^{-\alpha_2} - \ldots - c_{\alpha_k} x^{-\alpha_k} + o(x^{-\alpha_k}) \tag{6.3.4}$$

as $x \to \infty$, $\alpha_1 < \alpha_2 < \ldots < \alpha_k$, $0 < \alpha_1 < 1$. Then, as $n \to \infty$, $z/n^{1/\alpha_1} \to \infty$,

$$\mathsf{P}\,\{X_1 + \ldots + X_n > z\} = n \cdot \sum_{j=1}^{k} c_{\alpha_j} z^{-\alpha_j} + o\,(nz^{-\alpha_k}) -$$
$$- \frac{n^2}{2} \frac{(1 - 2\alpha_1)\,\Gamma\,(1 - \alpha_1)^2}{\Gamma\,(2 - 2\alpha_1)} c_{\alpha_1}^2 z^{-2\alpha_1} + o\,(n^2 z^{-2\alpha_1}). \qquad (6.3.5)$$

Proof. Take an arbitrary $\varepsilon \in (0, 2/5)$. The probability $\mathsf{P}\,\{X_1 + \ldots + X_n > z\}$ is split into the sum of $P_0 = \mathsf{P}\,\{X_1 + \ldots + X_n > z,\ X_i \le \varepsilon z,\ 1 \le i \le n\}$ and $\mathsf{P}\left(\bigcup_{i=1}^{n} \{X_1 + \ldots + X_n > z,\ X_i > \varepsilon z\}\right)$. The latter probability, by the known formulas connecting the probability of a union of events with the probabilities of their intersections, is between

$$\sum_{i=1}^{n} \mathsf{P}\,\{X_1 + \ldots + X_n > z,\ X_i > \varepsilon z\} -$$
$$- \sum_{1 \le i < j \le n} \mathsf{P}\,\{X_1 + \ldots + X_n > z,\ X_i > \varepsilon z,\ X_j > \varepsilon z\} \qquad (6.3.6)$$

and the same with the following sum added:

$$\sum_{1 \le i < j < l \le n} \mathsf{P}\,\{X_1 + \ldots + X_n > z, X_i > \varepsilon z, X_j > \varepsilon z, X_l > \varepsilon z\}. \quad (6.3.7)$$

Denote the first sum in (6.3.6) by P_1, the second one, by P_2, the sum (6.3.7) by P_3. We have

$$\mathsf{P}\,\{X_1 + \ldots + X_n > z\} = P_0 + P_1 - P_2 + O\,(P_3). \qquad (6.3.8)$$

We estimate P_3:

$$P_3 \le C_n^3 \mathsf{P}\,\{X_1 > \varepsilon z,\ X_2 > \varepsilon z,\ X_3 > \varepsilon z\} \le n^3 \mathsf{P}\,\{X_i > \varepsilon z\}^3 \sim$$
$$\sim (nc_{\alpha_1}(\varepsilon z)^{-\alpha_1})^3 = O\,(n^3 z^{-3\alpha_1}) = o\,(n^2 z^{-2\alpha_1}).$$

The probability P_0 can be represented in the form $\mathsf{P}\,\{\xi^{n,z} \in A^0\}$, where the set A^0 consists of all functions $x \in \boldsymbol{D}_0$, $x\,(1) > 1$, having no jumps of size exceeding ε. We easily find

$$\rho_{0,1}\,(A^0, B_0^2) = (1 - 2\varepsilon)/5 > 0;$$

by Lemma 6.2.1,

$$P_0 = \mathsf{P}\,\{\xi^{n,z} \in A^0\} \le \mathsf{P}\,\{\rho_{0,1}\,(\xi^{n,z},\ B_0^2) \ge (1 - 2\varepsilon)/5\} = O\,(n^3 z^{-3\alpha_1}).$$

The sum P_2 also, up to infinitesimals of higher order, coincides with the probability that $\xi^{n,z}$ hits a certain set, namely, the set A^2 consisting of all functions $x \in \boldsymbol{D}_0$, $x(1) > 1$, having at least two jumps exceeding ε. Indeed, $\xi^{n,z} \in A^2$ means that at least two of the events $\{X_1 + \dots + X_n > z,\ X_i > \varepsilon z\}$ occur, and it is easily proved that

$$P_2 - 2P_3 \le \mathsf{P}\{\xi^{n,z} \in A^2\} \le P_2,$$

so that

$$P_2 = \mathsf{P}\{\xi^{n,z} \in A^2\} + O(n^3 z^{-3\alpha_1}).$$

To the probability $\mathsf{P}\{\xi^{n,z} \in A^2\}$ we apply Theorem 6.2.3 with $g(n, z) = nz^{-\alpha_1}$ and the measure $\lambda_{t,y}(dx)$ given by formula (6.3.3) with $\alpha = \alpha_1$. We find $\rho_{0,1}(A^2, B_0^1) = \varepsilon/2 > 0$. For $0 < \delta < \varepsilon/2$ the set $A^2_{\pm\delta}$ consists of functions $x \in \boldsymbol{D}_0$, $x(1) > 1 \mp \delta$, having at least two jumps exceeding $\varepsilon \mp 2\delta$. Then we proceed:

$$X_{0,2}^{-1}(A^2) = \{(t_1, x_1, t_2, x_2)\colon 0 < t_1 < t_2 \le 1, \quad x_1 > \varepsilon,\ x_2 - x_1 > \varepsilon,\ x_2 > 1\},$$

and the set $[X_{0,2}^{-1}(A^2_{\pm\delta})] \setminus (X_{0,2}^{-1}(A^2_{-\delta}))$, up to sets of smaller dimension, consists of quadruples (t_1, x_1, t_2, x_2), $0 \le t_1 \le t_2 \le 1$, with

$$x_1 \in [\varepsilon - 2\delta,\ \varepsilon + 2\delta],\ x_2 - x_1 \ge \varepsilon - 2\delta,\ x_2 \ge 1 - \delta$$

or

$$x_1 \ge \varepsilon - 2\delta,\ x_2 - x_1 \in [\varepsilon - 2\delta,\ \varepsilon + 2\delta],\ x_2 \ge 1 - \delta,$$

or else

$$x_1 \ge \varepsilon - 2\delta,\ x_2 - x_1 \ge \varepsilon - 2\delta,\ x_2 \in [1 - \delta,\ 1 + \delta].$$

The μ_0^2-measure of this set converges to 0 as $\delta \downarrow 0$, and

$$\mu_0^2(X_{0,2}^{-1}(A^2)) = \frac{c_{\alpha_1}^2}{2} \iint\limits_{\substack{x_1 > \varepsilon \\ x_2 - x_1 > \varepsilon \\ x_2 > 1}} d(-x_1^{-\alpha_1})\, d(-(x_2 - x_1)^{-\alpha_1}) =$$

$$= \frac{c_{\alpha_1}^2}{2} \left[\int_\varepsilon^{1-\varepsilon} (1-t)^{-\alpha_1} d(-t^{-\alpha_1}) + (1-\varepsilon)^{-\alpha_1} \varepsilon^{-\alpha_1} \right].$$

Thus

$$P_2 = \frac{c_{\alpha_1}^2}{2} C(\varepsilon) \cdot n^2 z^{-2\alpha_1} + o(n^2 z^{-2\alpha_1}), \tag{6.3.9}$$

where $C(\varepsilon)$ is the expression between brackets in the previous formula.

6.3.3. Asymptotics of P_1. End of proof

We will estimate the sum P_1 more precisely, using the expansion (6.3.4) and the already proved Theorem 6.3.1:

$$P_1 = n\mathsf{P}\{X_1 + \dots + X_n > z,\ X_n > \varepsilon z\} =$$

$$= n\left[\mathsf{P}\{X_1 + \dots + X_{n-1} > (1-\varepsilon)z\} \cdot \mathsf{P}\{X_n > \varepsilon z\} + \right.$$

$$\left. + \int_0^{(1-\varepsilon)z} dF_{X_1 + \dots + X_{n-1}}(x) \cdot \mathsf{P}\{X_n > z - x\}\right] =$$

$$= n\left[((n-1)\, c_{\alpha_1}((1-\varepsilon)z)^{-\alpha_1} + o(nz^{-\alpha_1})) \cdot (c_{\alpha_1}(\varepsilon z)^{-\alpha_1} + o(z^{-\alpha_1})) + \right.$$

$$\left. + \int_0^{(1-\varepsilon)z} dF_{X_1 + \dots + X_{n-1}}(x) \left(\sum_{j=1}^{k} c_{\alpha_j}(z-x)^{-\alpha_j} + o\left((z-x)^{-\alpha_k}\right)\right)\right].$$

Here, since $x \le (1-\varepsilon)z$, $z - x \ge \varepsilon z$, we can replace $o((z-x)^{-\alpha_k})$ by $o(z^{-\alpha_k})$, uniformly with respect to x. Making use of the equivalence $n - 1 \sim n$ as well, we rewrite the expression for P_1 as follows:

$$P_1 = c_{\alpha_1}^2 (1-\varepsilon)^{-\alpha_1} \varepsilon^{-\alpha_1} n^2 z^{-2\alpha_1} + o(n^2 z^{-2\alpha_1}) +$$

$$+ n \sum_{j=1}^{k} c_{\alpha_j} \int_0^{(1-\varepsilon)z} (z-x)^{-\alpha_j}\, dF_{X_1 + \dots X_{n-1}}(x). \tag{6.3.10}$$

Rewrite the j-th integral in the form

$$z^{-\alpha_j}\left[\int_0^{(1-\varepsilon)z} 1\, dF_{X_1 + \dots + X_{n-1}}(x) + \right.$$

$$\left. + \int_0^{(1-\varepsilon)z} ((1 - xz^{-1})^{-\alpha_j} - 1)\, dF_{X_1 + \dots + X_{n-1}}(x)\right]. \tag{6.3.11}$$

The first integral is equal to

$$1 - \mathsf{P}\{X_1 + \dots + X_{n-1} > (1-\varepsilon)z\} =$$

$$= 1 - (n-1)\, c_{\alpha_1} ((1-\varepsilon) z)^{-\alpha_1} + o(nz^{-\alpha_1}); \tag{6.3.12}$$

let us prove that the second one is equal to

$$c_{\alpha_1} D_{\alpha_j}(\varepsilon)\,(n-1)\, z^{-\alpha_1} + o(nz^{-\alpha_1}), \tag{6.3.13}$$

where

$$D_{\alpha_j}(\varepsilon) = \int_0^{1-\varepsilon} ((1-t)^{-\alpha_j} - 1)\, d(-t^{-\alpha_1})$$

(the integral converges since $\alpha_1 < 1$).

It is sufficient to prove that for any $\kappa > 0$ the second integral in (6.3.11) differs, for sufficiently large n and $z/n^{1/\alpha_1}$, from $c_{\alpha_1} D_{\alpha_j}(\varepsilon)\,(n-1)\, z^{-\alpha_1}$ by at most $\kappa(n-1)\, z^{-\alpha_1}$. By Theorem 6.3.1,

$$F_{X_1+\dots+X_{n-1}}(x) = 1 - (n-1)\, c_{\alpha_1} x^{-\alpha_1} + o((n-1)\, x^{-\alpha_1})$$

as $n \to \infty$, $x/n^{1/\alpha_1} \to \infty$; so there exist n_0 and A such that for $n \geq n_0$, $x/n^{1/\alpha_1} \geq A$,

$$| F_{X_1+\dots+X_{n-1}}(x) - 1 + (n-1)\, c_{\alpha_1} x^{-\alpha_1} | \leq \frac{\kappa}{K}(n-1)\, x^{-\alpha_1}, \tag{6.3.14}$$

where K is a large positive constant. For sufficiently large n we split the second integral in in (6.3.11) into two: from 0 to An^{1/α_1} and from An^{1/α_1} to $(1-\varepsilon)\, z$. The first integral is at most

$$\max_{0 \leq x \leq An^{1/\alpha_1}} ((1 - xz^{-1})^{-\alpha_j} - 1) \sim \alpha_j A n^{1/\alpha_1} z^{-1} = \alpha_j A\, (nz^{-\alpha_1})^{1/\alpha_1} = o(nz^{-\alpha_1}).$$

In the second integral we use integration by parts:

$$\int_{An^{1/\alpha_1}}^{(1-\varepsilon)z} ((1 - xz^{-1})^{-\alpha_j} - 1)\, d\,[F_{X_1+\dots+X_{n-1}}(x) - 1] =$$

$$= -(\varepsilon^{-\alpha_j} - 1)\,[1 - F_{X_1+\dots+X_{n-1}}((1-\varepsilon)\, z)] +$$

$$+ \left(\left(1 - An^{1/\alpha_1} z^{-1}\right)^{-\alpha_j} - 1\right)\left[1 - F_{X_1+\dots+X_{n-1}}\left(An^{1/\alpha_1}\right)\right] +$$

$$+ \int_{An^{1/\alpha_1}}^{(1-\varepsilon)z} [1 - F_{X_1+\dots+X_{n-1}}(x)]\, d\,((1 - xz^{-1})^{-\alpha_j} - 1).$$

By (6.3.14) this expression lies within the bounds

$$(n-1)\Big\{-(\varepsilon^{-\alpha_j}-1)\,(c_{\alpha_1}\pm\kappa/K)\,((1-\varepsilon)\,z)^{-\alpha_1}+$$

$$+\,(c_{\alpha_1}\mp\kappa/K)\Big[\Big(\big(1-An^{1/\alpha_1}z^{-1}\big)^{-\alpha_j}-1\Big)\big(An^{1/\alpha_1}\big)^{-\alpha_1}+$$

$$+\int_{An^{1/\alpha_1}}^{(1-\varepsilon)\,z} x^{-\alpha_1}\,d\,((1-xz^{-1})^{-\alpha_j}-1\Big]\Big\}=$$

$$=(n-1)\Big\{(c_{\alpha_1}\mp\kappa/K)\Big[\;-x^{-\alpha_1}\,((1-xz^{-1})^{-\alpha_j}-1)\Big|_{An^{1/\alpha_1}}^{(1-\varepsilon)\,z}+$$

$$+\int_{An^{1/\alpha_1}}^{(1-\varepsilon)\,z} x^{-\alpha_1}\,d\,((1-xz^{-1})^{-\alpha_j}-1\Big]\mp$$

$$\mp\;2\,(\varepsilon^{-\alpha_j}-1)\,((1-\varepsilon)\,z)^{-\alpha_1}\cdot\kappa/K\Big\}.$$

Integrating by parts in the opposite direction we reduce the expression between brackets to

$$\int_{An^{1/\alpha_1}}^{(1-\varepsilon)\,z}((1-xz^{-1})^{-\alpha_j}-1)\,d\,(-x^{-\alpha_1})=$$

$$=z^{-\alpha_1}\int_{An^{1/\alpha_1}z-1}^{1-\varepsilon}((1-t)^{-\alpha_j}-1)\,d\,(-t^{-\alpha_1})=$$

$$=z^{-\alpha_1}\,(D_{\alpha_j}(\varepsilon)+o\,(1))\;(\le D_{\alpha_j}(\varepsilon)\,z^{-\alpha_1}).$$

Finally we obtain that the second integral in (6.3.11) is between

$$(n-1)\,z^{-\alpha_1}\Big[c_{\alpha_1}\mp\frac{\kappa}{K}\,(D_{\alpha_j}(\varepsilon)+2\,(\varepsilon^{-\alpha_j}-1)\,(1-\varepsilon)^{-\alpha_1})\mp o\,(1)\Big].$$

Choosing $K>D_{\alpha_j}(\varepsilon)+2\,(\varepsilon^{-\alpha_j}-1)\,(1-\varepsilon)^{-\alpha_1}$, we obtain the estimate required.

Taking the formulas (6.3.10) - (6.3.13) together and using the equivalence $n - 1 \sim n$, we obtain:

$$P_1 = c^2_{\alpha_1} (1-\varepsilon)^{-\alpha_1} \varepsilon^{-\alpha_1} n^2 z^{-2\alpha_1} + o\,(n^2 z^{-2\alpha_1}) + \sum_{j=1}^{k} c_{\alpha_j} n z^{-\alpha_j} + o\,(nz^{-\alpha_k}) +$$

$$+ \sum_{j=1}^{k} \left[c_{\alpha_j} c_{\alpha_1} \left(D_{\alpha_j}(\varepsilon) - (1-\varepsilon)^{-\alpha_1}\right) n^2 z^{-\alpha_1 - \alpha_j} + o\,(n^2 z^{-\alpha_1 - \alpha_j}) \right]. \tag{6.3.15}$$

The terms $o\,(n^2 z^{-\alpha_1 - \alpha_j})$ and those with $n^2 z^{-\alpha_1 - \alpha_j}$ for $j > 1$ are combined into $o\,(n^2 z^{-2\alpha_1})$.

Substituting the expressions (6.3.15) for P_1, (6.3.9) for P_2 and the estimates for P_0 and P_3 into (6.3.8), we obtain:

$$\mathsf{P}\,\{X_1 + \ldots + X_n > z\} = \sum_{j=1}^{k} c_{\alpha_j} n z^{-\alpha_j} + o\,(nz^{-\alpha_k}) +$$

$$+ c^2_{\alpha_1} n^2 z^{-2\alpha_1} \left[\int_0^{1-\varepsilon} ((1-t)^{-\alpha_1} - 1)\, d\,(-t^{-\alpha_1}) - (1-\varepsilon)^{-\alpha_1} + \right.$$

$$+ (1-\varepsilon)^{-\alpha_1} \varepsilon^{-\alpha_1} - \frac{1}{2} \int_{\varepsilon}^{1-\varepsilon} (1-t)^{-\alpha_1}\, d\,(-t^{-\alpha_1}) +$$

$$\left. + \frac{1}{2} (1-\varepsilon)^{-\alpha_1} \varepsilon^{-\alpha_1} \right] + o\,(n^2 z^{-2\alpha_1}).$$

The expression between brackets, naturally, cannot depend on ε. Differentation with respect to ε confirms this. This expression can be found by limit transition as $\varepsilon \downarrow 0$; it proves to be equal to

$$\frac{1}{2} \left[\int_0^1 ((1-t)^{-\alpha_1} - 1)\, d\,(-t^{-\alpha_1}) - 1 \right] = -\frac{1}{2} \frac{(1-2\alpha_1)\,\Gamma\,(1-\alpha_1)^2}{\Gamma\,(2-2\alpha_1)}.$$

This proves (6.3.5). ◊

REFERENCES

Anderson R.F., Orey S.

1. *Small random perturbations of dynamical systems with reflecting boundary.* Nagoya Math. J., 1976, Vol. 60, pp. 189-216.

Azencott R., Ruget G.

1. *Mélanges d'équations différentielles et grands écarts à la loi des grands nombres.* Z. Wahrscheinlichkeitstheorie und verw. Gebiete, 1977, Vol. 38, pp. 1-54.

Bahadur R.R.

1. *On the asymptotic efficiency of tests and estimates.* Sankhyā, 1960, Vol. 22, N. 3-4, pp. 229-252.

Billingsley P.

1. *Convergence of probability measures.* New York etc.: John Wiley & Sons, Inc., 1968.

Borovkov A.A.

1. *Boundary-value problems for random walks and large deviations in function spaces.* Teor. Veroyatn. i Primen., 1967, Vol. 12, N. 4, pp. 635-654.

Borovkov A.A., Mogul'skii, A.A.

1. *On probabilities of large deviations in topological spaces. I*: Sibirsk. Mat. Zh., 1978, Vol. 19, N. 5, pp. 988-1004. *II*: 1980, Vol. 21, N. 1, pp. 12-26.

Bourbaki N.

1. *Elements of Mathematics.* Vol. 3, General Topology, Reading (MA): Addison-Wesley, 1977, translated from the French.

Cramér H.

1. *Sur un nouveau théorème limite de la théorie des probabilités.* Act. Sci. et Ind., 1938, Hermann, Paris, f. 736.

Donsker M. D., Varadhan, S.R.S.

1. *Aymptotic evaluation of certain Markov process expectations for large time. I*: Comm. Pure Appl. Math., 1975, Vol. 28, N. 1, pp. 1-47; *II*: 1975, Vol. 28, N. 2, pp. 279-301; *III*: 1976, Vol. 29, N. 4, pp. 389-461.
2. *Asymptotics for the Wiener sausage.* Comm. Pure Appl. Math., 1975, Vol. 28, N. 4, pp. 525-565.

Dzhakhangirova A.D., Nagaev, A.V.

1. *The multidimensional central limit theorem, taking into account large deviations.* In: Random processes and related problems, Tashkent, Fan, 1971, Vol. 2, pp. 25-35 (in Russian).

Dubrovskii, V.N.

1. *An asymptotic formula of Laplace type for discontinuous Markov processes.* Teor. Veroyatn. i Primen., 1976, Vol. 21, N. 1, pp. 219-222 (in Russian).

2. *Exact asymptotic formulas of Laplace type for Markov processes*. Soviet Math. Dokl., 1976, Vol. 17, pp. 223-227.

Dynkin, E.B.

1. *Die Grundlagen der Theorie der Markoffschen Prozesse*. Berlin etc.: Springer, 1961 (translated from Russian).

Fortus M.I.

1. *A uniform limit theorem for distributions which are attracted to a stable law with index less than one*. Theor. Probab. Appl., 1957, Vol. 2, pp. 478-479.

Freidlin M.I.

1. *The action functional for a class of stochastic processes*. Theor. Probab. Appl., 1972, Vol. 17, pp. 511-515.
2. *On stability of highly reliable systems*. Theor. Probab. Appl., 1975, Vol. 20, pp. 572-583.
3. *Fluctuations in dynamical systems with averaging*. Soviet Math. Dokl., 1976, Vol. 17, pp. 104-108.
4. *Sublimiting distributions and stabilizations of solutions of parabolic equations with a small parameter*. Soviet Math. Dokl., 1977, Vol. 18, pp. 1114-1118.
5. *The averaging principle and theorems on large deviations*. Russian Math. Surveys, 1978, Vol. 33, pp. 117-176.

Freidlin M. I., Ventcel' A. D.

1. *Small random perturbations of a dynamical system with a stable equilibrium position*. Soviet Math. Dokl., 1969, Vol. 10, pp. 886-890.
2. *On the limit behaviour of an invariant measure under small random perturbations of a dynamical system*. Soviet Math. Dokl., 1969, Vol. 10, pp. 1047-1051.
3. *On motion of diffusing particles against a current*. UMN, 1969, Vol. 24, N. 5, pp. 229-230 (in Russian).
4. *On small random perburtations of dynamical systems*. Russian Math. Surveys, 1970, Vol. 25, pp. 1-55.
5. *Some problems concerning stability under small random perburtations*. Theor. Probab. Appl., 1972, Vol. 17, pp. 269-283.
6. *Random perburtations of dynamical systems*. New York etc.: Springer, 1984 (translated from the Russian).

Friedman A.

1. *Small random perturbations of dynamical systems and applications to parabolic equations*. Indiana Univ. Math. J., 1974, Vol. 24, N. 6, pp. 533-553. Erratum to this paper: ibid., 1975, Vol. 24, N. 9.
2. *The asymptotic behaviour of the first real eigenvalue of the second order elliptic operator with a small parameter in the highest derivatives*. Indiana Univ. Math. J., 1973, Vol. 22, N. 10, pp. 1005-1015.

Gärtner, Yu.
1. *On large deviations from the invariant measure.* Theor. Probab. Appl., 1977, Vol. 22, pp. 24-39.

Gikhman, I.I., Skorokhod, A.V.
1. *Theory of random processes.* New York etc.: Springer, 1977 (translated from the Russian).

Godovan'chuk, V.V.
1. *Probabilities of large deviations for sums of independent random variables attracted to a stable law.* Theor. Probab. Appl., 1978, Vol. 23, pp. 602-608.
2. *Asymptotic probabilities of large deviations due to large jumps of a Markov process.* Theor. Probab. Appl., 1981, Vol. 26, pp. 314-327.

Grigelionis B.
1. *On the structure of densities of measures corresponding to random processes.* Litov. Mat. Sb., 19173, Vol. 13, N. 1, pp. 71-78 (in Russian).

Grin', A.G.
1. *On perturbations of dynamical systems by regular Gaussian processes.* Teor. Veroyatn. i Primen., 1975, Vol. 20, N. 2, pp. 456-457 (in Russian).

Halmos P.R.
1. *Measure theory.* New York, 1950, v. Nostrand.

Heyde C.C.
1. *On large deviation probabilities in the case of attraction to a non-normal stable law.* Sankhyā, 1968, Vol. A 30, N. 3, pp. 253-258.

Ibragimov I.A., Linnik Yu.V.
1. *Independent and stationary sequences of variables.* Wolters-Noordhoff, Groningen, 1971 (translated from Russian).

Kifer Yu.I.
1. *On the asymptotics of the transition density of processes with small diffusion.* Theor. Probab. Appl., 1976, Vol. 21, pp. 513-522.

Komatsu T.
1. *Markov processes associated with certain integro-differential equations.* Osaka J. Math., 1973, Vol. 10, pp. 271-303.

Korostelev A.P.
1. *A convergence criterion of continuous stochastic approximation procedures.* Theor. Probab. Appl., 1977, Vol. 22, 584-591.
2. *Damped perturbations of dynamical systems and convergence conditions for stochastic procedures.* Theor. Probab. Appl., 1979, Vol. 24, pp. 302-321.

Kunita H., Watanabe S.
1. *On square integrable martingales.* Nagoya Math. J., 1967, Vol. 30, pp. 209-245.

Lepeltier J.P., Marchal B.
1. *Problème des martingales et équations différentielles stochastiques associées à un opérateur intégro-différentiel.* Ann. Inst. H. Poincaré, 1976, Vol. B 12, N. 1, pp. 43-103.
Liptser R. S., Shiryayev A.N.
1. *Statistics of random processes. Vol. II, Applications.* Berlin etc.: Springer, 1978 (translated from the Russian).
Maslov V.P.
1. *Global asymptotic expansion of probabilities of large deviations and its relation with quasi-classical asymptotics.* UMN, 1979, Vol. 34, N. 5, pp. 213 (in Russian).
2. *Global exponential asymptotics of solutions of tunnel-type equations.* Russian Math. Surveys, 1981, Vol. 24, pp. 655-659.
Meyer P.-A.
1. *Un cours sur les intégrales stochastiques.* In: Lecture Notes in Math., Vol. 511, Berlin etc.: Springer, 1976, pp. 245-300.
Mogul'skii A.A.
1. *Large deviations for trajectories of multi-dimensional random walks.* Theor. Probab. Appl., 1976, Vol. 21, pp. 300-315.
2. *Probabilities of large deviations for trajectories of random walks.* In: Limit theorems for sums of random variables, Novosibirsk, Nauka, 1984, pp. 93-124 (in Russian).
Molchanov S.A.
1. *Diffusion processes and Riemannian geometry.* Russian Math. Surveys, 1975, Vol. 30, pp. 1-63.
Nagaev A.V.
1. *Probabilities of large deviations of sums of independent random variables.* Thesis. Tashkent, 1970 (in Russian).
Nguyen Viet Phu
1. *On a problem of stability under small perturbations.* Moscow Univ. Math. Bull., 1974, Vol. 29, pp. 5-11.
2. *Small Gaussian perturbations and Euler's equation.* Moscow Univ. Math. Bull., 1974, Vol. 29, pp. 52-57.
Pinelis F.
1. *A problem on large deviations in a space of trajectories.* Theor. Probab. Appl., 1981, Vol. 26, pp. 69-84.
Rockafellar R.T.
1. *Convex analysis.* Princeton, NJ: Princeton University Press, 1970.

Sanov I.N.

1. *On probabilities of large deviations of random variables*. Mat. Sb., 1957, Vol. 42 (84), N. 1, pp. 11-44 (in Russian).

Schilder M.

1. *Some asymptotic formulas for Wiener integrals*. TAMS, 1966, Vol. 125, N. 1, pp. 63-85.

Stroock D.W.

1. *Diffusion processes with Lévy generators*. Z. Wahrscheinlichkeitstheorie und verw. Gebiete, 1975, Vol. 32, pp. 209-241.

Stroock D.W., Varadhan, S.R.S.

1. *Diffusion processes with continuous coefficients: I*. Comm. Pure Appl. Math., 1969, Vol. 22, N. 3, pp. 345-400; *II*, 1969, Vol. 22, N. 4, pp. 479-530.
2. *Multidimensional diffusion process*. Berlin etc.: Springer, 1979.

Tkachuk S.G.

1. *A theorem on large deviations in the case of distributions with regularly varying tails*. In: Random processes and statistic inferences, Tashkent, Fan, 1975, Vol. 5, pp. 164-174 (in Russian).

Varadhan S.R.S.

1. *Asymptotic probabilities and differential equations*. Comm. Pure Appl. Math., 1969, Vol. 19, N. 3, pp. 261-286.
2. *On the behaviour of the fundamental solution of the heat equation with variable coefficients*. Comm. Pure Appl. Math., 1967, Vol. 20, N. 2, pp. 431-455.
3. *Diffusion processes in a small time interval*. Comm. Pure Appl. Math., 1967, Vol. 20, N. 4, pp. 659-685.

Vinogradov, V.V.

1. *Asymptotic expansions in limit theorems on large deviations*. In: XXII All-Union Sci. Student Conf. Mat., Novosibirsk, NGU, 1984, pp. 17-21 (in Russian).

Wentzell, A.D.

1. *On the asymptotic behaviour of the greatest eigenvalue of a second order elliptic differential operator with a small parameter in the highest derivatives*. Soviet Math. Dokl., 1972, Vol. 13, pp. 13-18.
2. *On the asymptotics of eigenvalues of matrices with elements of order* $\exp\{-V_{ij}/2\varepsilon^2\}$. Soviet Math. Dokl., 1972, Vol. 13, pp. 65-68.
3. *Theorems on the action functional for Gaussian random functions*. Theor. Probab. Appl., 1972, Vol. 17, pp. 515-517.
4. *Limit theorems on large deviations for random processes*. Theor. Probab. Appl., 1973, Vol. 18, pp. 817-821.
5. *Limit theorems on large deviations for Markov processes*. In: Internat. Conf.

Probab. Theory and Math. Stat., Vilnius, 1973, Vol. 18, pp. 133-136 (in Russian).

6. *On the asymptotic behaviour of the first eigenvalues of a second order differential operator with small parameter in higher derivatives.* Theor. Probab. Appl., 1975, Vol. 20, pp. 599-602.
7. *Rough limit theorems on large deviations for Markov processes.* Theor. Probab. Appl., 1976, Vol. 21, pp. 227-242, pp. 499-512; Vol. 24, 1979, pp. 675-692; Vol. 27, 1982, pp. 215-234.

   [see also Freidlin]

Zhivoglyadova, L.V., Freidlin, M.I.

1. *Boundary problems with a small parameter for a diffusion process with reflection.* UMN, 1976, Vol. 31, N. 5, pp. 241-242 (in Russian).

SUBJECT INDEX

Some notions are used regularly throughout the whole text (e.g. the action functional for a family of stochastic processes is used throughout Chapters 3-5). These usages are not followed up in the Index. We also do not give the pages in such instances of concept usage, as can be seen from the Table of Contents.

9 780792 301431